Lecture Notes in Computer Science 10720

Commenced Publication in 1973
Founding and Former Series Editors:
Gerhard Goos, Juris Hartmanis, and Jan van Leeuwen

Abdelkader Hameurlain · Josef Küng
Roland Wagner · Tran Khanh Dang
Nam Thoai (Eds.)

Transactions on Large-Scale Data- and Knowledge-Centered Systems XXXVI

Special Issue on Data
and Security Engineering

 Springer

Editors-in-Chief

Abdelkader Hameurlain
IRIT, Paul Sabatier University
Toulouse
France

Roland Wagner
FAW, University of Linz
Linz
Austria

Josef Küng
FAW, University of Linz
Linz
Austria

Guest Editors

Tran Khanh Dang
Ho Chi Minh City University of Technology
Ho Chi Minh City
Vietnam

Nam Thoai
Ho Chi Minh City University of Technology
Ho Chi Minh City
Vietnam

ISSN 0302-9743   ISSN 1611-3349   (electronic)
Lecture Notes in Computer Science
ISSN 1869-1994   ISSN 2510-4942   (electronic)
Transactions on Large-Scale Data- and Knowledge-Centered Systems
ISBN 978-3-662-56265-9   ISBN 978-3-662-56266-6   (eBook)
https://doi.org/10.1007/978-3-662-56266-6

Library of Congress Control Number: 2017960854

This Springer imprint is published by Springer Nature
The registered company is Springer-Verlag GmbH, DE
The registered company address is: Heidelberger Platz 3, 14197 Berlin, Germany

Preface

The 3rd International Conference on Future Data and Security Engineering (FDSE) and the 10th International Conference on Advanced Computing and Applications (ACOMP) were held in Can Tho City, Vietnam, during November 23–25, 2016. FDSE is an annual international premier forum designed for researchers and practitioners interested in state-of-the-art and state-of-the-practice activities in data, information, knowledge, and security engineering to explore cutting-edge ideas, present and exchange their research results and advanced data-intensive applications, as well as to discuss emerging issues on data, information, knowledge, and security engineering. ACOMP annual events focus on advanced topics in computer science and engineering. More specifically, ACOMP 2016 solicited papers about security, information systems, software engineering, enterprise engineering, embedded systems, high performance computing, image processing, visualization, and other related fields.

We encouraged the submission of both original research contributions and industry papers. FDSE 2016 received 115 submissions and, after a careful review process, only 29 papers (27 full and 2 short ones) were selected for presentation. The call for papers of ACOMP 2016 resulted in the submissions of a total of 61 papers, out of which we selected only 22 papers for presentation, following a rigorous review process.

Among the great papers of both FDSE and ACOMP 2016, we selected 9 papers to invite the authors to revise, extend, and resubmit for publication in this special issue. At the end, only 8 extended papers were accepted. The main focus of this special issue is on advanced computing issues in data and security engineering, as well as their promising applications.

The big success of FDSE and ACOMP 2016, as well as this special issue of TLDKS, was the result of the efforts of many people, to whom we would like to express our gratitude. First, we would like to thank all authors who extended and submitted papers to this special issue. We would also like to thank the members of the committees and external reviewers for their timely reviewing and lively participation in the subsequent discussion in order to select the high-quality papers published in this issue. Finally, yet importantly, we thank Gabriela Wagner for her enthusiastic help and support during the whole process of preparation for this publication.

October 2017

Tran Khanh Dang
Nam Thoai

The original version of the book Frontmatter was revised: a subtitle has been added in the Title page. The erratum to the book Frontmatter is available at https://doi.org/10.1007/978-3-662-56266-6_9

Organization

Editorial Board

Viet Hung Nguyen   Bosch, Vietnam, and Trento University, Italy
Le Minh Sang Tran   Trento University, Italy
Minh Quang Tran   HCMC University of Technology, VNUHCM, Vietnam
Ngoc Thinh Tran   HCMC University of Technology, VNUHCM, Vietnam
Anh Truong   HCMC University of Technology, VNUHCM, Vietnam,
                        and Trento University, Italy
Hoang Tam Vo   IBM Research, Australia

Contents

Risk-Based Privacy-Aware Access Control for Threat Detection Systems

Nadia Metoui[1(✉)], Michele Bezzi[2], and Alessandro Armando[3]

[1] DISI, University of Trento, Trento, Italy
nadia.metoui@gmail.com
[2] SAP Labs France, Security Research, Sophia-Antipolis, France
[3] DIBRIS, University of Genova, Genoa, Italy

Abstract. Threat detection systems collect and analyze a large amount of security data logs for detecting potential attacks. Since log data from enterprise systems may contain sensitive and personal information access should be limited to the data relevant to the task at hand as mandated by data protection regulations. To this end, data need to be pre-processed (anonymized) to eliminate or obfuscate the sensitive information that is *not-strictly necessary* for the task. Additional security/accountability measures may be also applied to reduce the privacy risk, such as logging the access to the personal data or imposing deletion obligations. Anonymization reduces the privacy risk, but it should be carefully applied and balanced with utility requirements of the different phases of the process: a preliminary analysis may require fewer details than an in-depth investigation on a suspect set of logs. We propose a risk-based privacy-aware access control framework for threat detection systems, where each access request is evaluated by comparing the privacy-risk and the trustworthiness of the request. When the risk is too large compared to the trust level, the framework can apply adaptive adjustment strategies to decrease the risk (e.g., by selectively obfuscating the data) or to increase the trust level to perform a given task (e.g., imposing enforceable obligations to the user). We show how the framework can simultaneously address both the privacy and the utility requirements. The experimental results presented in the paper that the framework leads to meaningful results, and real-time performance, within an industrial threat detection solution.

Parts of this paper have been presented at the International Conference on Future Data and Security Engineering [28]. This paper extends [28] in a number of ways: the theoretical framework has been extended so to include, in particular, explicit *deny* to data access (cf. Sect. 3.1); the adjustment decision from the authorization standpoint and its effect on the resulting response is now described; a new section (Sect. 4) presenting an architecture for proposed risk-based and privacy-aware access control framework and the data request evaluation workflow using the use case as running example has been added; a sketch of the policies to be used to express the risk-based authorizations (Sect. 5) is given; finally, both the Introduction and Conclusions have been improved.

A. Hameurlain et al. (Eds.): TLDKS XXXVI, LNCS 10720, pp. 1–30, 2017.
https://doi.org/10.1007/978-3-662-56266-6_1

Keywords: Trust · Risk · Privacy · Utility
Privacy-preserving threat detection

1   Introduction

Due to the increasing complexity and variety of attacks, modern Threat Detection Systems (TDS) are becoming more sophisticated and data-intensive. They leverage the correlation of security events from several log files to detect and prevent cyber attacks [34,45]. This is typically done in two main steps: an automatic pattern or anomaly detection phase which highlights suspicious events followed by a detailed investigation carried out by a human expert who must decide whether the anomalous pattern corresponds to an actual attack. In this second phase, the expert must often inspect the raw data (log files) that triggered the alert.

Although the security investigation can constitute a legitimate purpose for their processing of the log data, whenever they contain sensitive or personal information (e.g., user ids, IP addresses, logins) access must be limited to the relevant data for the analysis at hand. To this aim, data must be pre-processed (anonymized) to obfuscate those elements that are not strictly necessary for the task at hand. In addition, other security/accountability measures may be applied to reduce the risk, such as logging the access to the personal data or imposing deletion obligations. Anonymization is often used to pre-process the data, removing sensitive information from log files and enabling further processing with minimal privacy risk. However, the application of anonymization techniques can deteriorate the quality or utility of the data. Although some analytics can still be run on anonymized log data [24], in many cases the anonymization affects the quality of results and, ultimately, decreases the ability to detect and react to cyber threats.

We propose a risk-based privacy-aware control framework for TDS that addresses the concerns described above by enabling access to the data only if the level of risk does not exceed a certain threshold (related to the *trust level* of the user) and by allowing for risk reduction (e.g., by anonymization) or trust enhancement access. The framework does not require an *a priori* risk mitigation measure, i.e. off-line, anonymization of the data sources. The automatic pattern detection phase uses the original data set, and anonymization is applied only if a subsequent, human-based analysis is needed on the resulting data.

The risk level of each data request is dynamically evaluated by the access control decision point based on several parameters (e.g., context, role, and trustworthiness of the requester) and (if needed) anonymization is applied on the specific resulting data set. In the paper we focus on *re-identification* risk and, following common practice, we use k-anonymity as risk metrics. However, the approach is not bound to these choices and it can be readily adapted to alternative metrics (e.g., ℓ-diversity, t-closeness, and differential privacy).

The proposed approach has a number of advantages:

- it limits the impact on the utility, since we apply the anonymization only after running the pattern detection on the original data, and we adapt the anonymization strategy to the specific pattern;
- it provides a simple framework to address the, often conflicting, privacy and utility requirements;
- it is based on concepts as trust and risk, which have an intuitive meaning in the business world;
- it can be easily configured so to support the best trade-off between security and privacy according to priorities of the organization: risk and trust levels can be easily changed, the adjustment strategy can be configured to optimize utility or performance goals, etc.);
- it can be realized using a *declarative* policy language with a number of advantages including usability, flexibility, and scalability. As we will see in Sect. 5 the architecture and the policy structure we propose can be readily implemented as an extension of a well-known declarative authorization language (XACML).

To evaluate the effectiveness of the proposed approach, we have developed a prototype implementation and we experimentally evaluated it by running a number of threat detection patterns based on the SAP Enterprise Threat Detection (ETD) solution. The results obtained (reported in Sect. 6.6) show that the model meets the utility and performance requirements of a realistic use-case.

Structure of the Paper. In the next section, we introduce a TDS use case which we use to illustrate the main features of our risk-based privacy-aware approach. In Sect. 3 we present our framework its application to the proposed use case. In Sect. 4 we describe the architecture of our framework and the risk-based privacy-aware request evaluation workflow. In Sect. 5 we describe how risk-based authorizations can be expressed through attribute-based policies and we provide some policy examples. Section 6 discusses the results of an experimental evaluation of the proposed approach in terms of performance, scalability, and data utility (after anonymization). Lastly, we discuss the related work in Sect. 7 and we conclude in Sect. 8 with some final remarks.

2   Use Case

Modern intrusion detection systems at application-level (called Threat Detection System, TDS, herein)[1], collect security information on the application stack and correlate it with context information to detect potential threats. Usually, a TDS first collects application-level log files from various sources, enriches the

---

[1] We refer to these systems as TDS, to distinguish them from network-level intrusion detection systems (often called IDS or SIEM). We base our description on the SAP Enterprise Threat Detection, but the analysis can be applied to other solutions, including IDS. For a comparison between application- and network-level intrusion detection systems, see [22].

data gathered from logs with contextual information (e.g., time and location), and finally stores the resulting data in a database. The events data are then automatically analyzed on a periodic basis against pre-defined threat patterns to detect potential anomalies and attacks. A pattern represents a combination of suspicious log events that could indicate a threat. Often it is defined as a set of filters applied to the event database and compared with some thresholds. If the threshold is exceeded, then an alert is triggered. For instance, the ensemble of events indicating a *Failed Login* initiated by the same source (e.g., Terminal) may indicate that a *Brute Force Attack* is underway if the number of attempts exceeds, say, 20 attempts in less than 10 min. When an alert is raised a human operator is asked to step in order to evaluate if the alert corresponds to an actual threat and when this is the case to undertake appropriate countermeasures. To carry out his task, the operator may require access to the details of the data that triggered the alert. The operator should be granted access to sensitive data if this is strictly necessary to carry out her task and the severity of the problem justifies it.

Figure 1 illustrates the architecture of the system as well as the different users involved in the process.

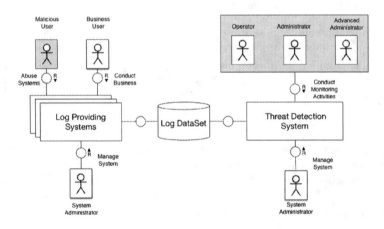

Fig. 1. Business roles and system landscape

Table 1 provides an example of user roles in the TDS and the corresponding access authorizations required to execute their tasks.

Although log files may contain personal information (e.g. names, IP addresses) investigation can constitute a legitimate purpose for their processing. Yet access to sensitive data should be done according to the data minimization principle, i.e. that access to personal information should be limited to what is directly relevant and necessary to accomplish the specified purpose. This is usually achieved in TDS by carrying out some (pseudo-)anonymization before analyzing the event data, such as replacing real user name or IDs with pseudonyms.

Table 1. Roles

Operator	Classify alerts and report patterns anomalies His/Her tasks require access to pattern detection results (events/log data related to the suspicious pattern) in case of alerts
Administrator	Has all *Operator* tasks and privileges. They can also Investigate alerts, Create or Reconfigure patterns. He/She should have access the detection results and events data related to the patterns
Advanced administrator	Has all *Administrator* tasks and privileges. Can also grant exceptional access to the data by attributing higher trust level to an *Operator* or an *Administrator*

Still, with the increasing variety and complexity of collected log files, a full anonymization of the log dataset before processing could, on one hand, provide a good privacy protection, but also significantly impact the performance of the system, both in terms of the *utility* (the quality of results of the pattern detection phase, or the information available to the operator for the manual inspection) and processing time (anonymization on large data set could be time-consuming, and on data stream re-run regularly)

To address this challenge, a more dynamic approach is needed: instead of anonymizing the complete event data-base beforehand, whenever a user performs an operation accessing event tables, we have to apply specific anonymization methods which reduce the privacy risk but preserving the most relevant information for that operation. In practice, the anonymization process should be customized for each operation (to preserve the information useful for completing the task) and for each type of users, which can have different levels of access to the data. In the next section, we will propose a framework that to realize this scenario.

3 The Model

In this section, we provide a general description of our risk-based privacy aware access control model, based on a previous model we introduced in [3], and we explain how it can be adapted to the use case described in Sect. 2.

3.1 Risk-Based Approach

The framework evaluates access decisions using the trust and risk values associated with the access request. An *access request* issued by subject u to carry out action a (e.g., read or write) on resource obj (e.g., a file) in context C is modeled as a quadruple $req = (u, a, obj, C)$. For instance, a request issued by user Alice to read file data.html in a security emergency context Alert1 is represented by $req_0 = (\text{Alice}, \text{Read}, \text{data.html}, \text{Alert1})$. Let Π be the access control policy of the organization. We write $\Pi(req) = $ granted to denote that req is granted access by Π. Policy Π can be extended so to take into account the risk $R(req)$ and the trustworthiness $T(req)$ of the request req by defining:

$$Auth_\Pi(req) = \begin{cases} \textbf{deny} & \text{if } \Pi(req) \neq \text{granted} \\ \sigma(req) & \text{if } T(req) - R(req) < 0 \\ \textbf{grant} & \text{otherwise} \end{cases} \qquad (1)$$

If the authorization is not *denied* by Π, the request is evaluated by comparing the risk $R(req)$ with the trustworthiness $T(req)$. $T(req)$ plays the role of a risk threshold (in practice, the maximum amount of risk that a requester can take in a certain context). If $T \geq R$ access is granted, otherwise it cannot be granted *as is*. In the latter case, instead of denying access, the system may identify and propose an *adjustment strategy* σ whose application meets the condition $T \geq R$. Adjustment strategies can be either *(i) risk mitigating*, i.e. mitigation strategies whose application decreases the risk R or *(ii) trust enhancing*, i.e. mitigation strategies whose application increases the trustworthiness T. Examples of risk mitigating strategies are data anonymization and the imposition of obligations of the handling of data. An example of trust enhancing strategies are the (temporary) privilege escalation and provision of an additional, stronger proofs of identity (e.g. a two-factor authentication) [3]. It must be noted that adjustment strategies normally bring some *negative* side effects. For example, anonymization degrades data quality and this may affect its utility. Privilege escalation can increase the complexity of the security governance. Thus to identify the best adjustment strategy it is necessary to strike a balance between the advantages brought by the application of the strategy and the associated side effects.

If we focus on data access and privacy risk (as the use case in Sect. 2) and limit the adjustment strategies to anonymization, then we can find an optimal anonymization strategy $\hat\sigma$, among all the possible anonymization strategies, that allows for data access reducing risk (so fulfilling Eq. 1) and, at the same time, maximizing the utility after application of the strategy. This can be expressed as a utility-privacy optimization problem:

$$\hat\sigma = \arg\max_\sigma U(obj')$$

$$\text{s.t. } req' = \sigma(req) \text{ and } R(req') \leq T(req')$$

where obj' denotes the resource in the request generated by the adjustment strategy, i.e. $req' = (u', a', obj', C')$.

In practice (as we will see in Sect. 3.2), the number of mitigation strategies is often very limited. The optimization problem is therefore reduced to testing a small set of anonymization strategies and estimating (either on the basis of numerical thresholds or expert assessment) if the utility of the result is sufficient for the business task under consideration. If this is not the case, trust enhancement mechanism can be triggered.

In the next subsections, we will show how trust and risk can be modeled, with a focus on the application to a TDS.

3.2   Privacy Enhancing Approach

Risk Model. Risk is generally expressed in terms the likelihood of the occurrence of certain (negative) events system and the (negative) impact of these events [18]. In this paper, we focus the risk associated with privacy breaches in information systems. Privacy breaches are often associated with *individual identifiability*, used in most data protection privacy laws (e.g., the EU data protection directive [32], the Health Insurance Portability and Accountability Act (HIPAA) [38]). To prevent individual identifiability the regulation requires that disclosed information (alone or in combination with reasonably available information from other sources or auxiliary information [31]) should not allow an intruder to identify individuals in a dataset (identity disclosure) or to learn private/sensitive information about individuals (attribute disclosure) with a very high probability or confidence (see [40,43]).

To assess the privacy risk (when releasing a given dataset) various privacy metrics have been proposed in the literature (see [8,14] for a review). The most popular metric is k-anonymity [36][2]. In the k-anonymity approach, attributes (or columns) in a dataset are classified as:

- *Identifiers*, i.e. attributes that can uniquely identify individuals e.g., full name, social security number passport number;
- *Quasi-identifiers (QIs) or key attributes*, i.e. attributes that, when combined, can be used to identify individuals, e.g., age, job function, postal code;
- *Sensitive attributes*, i.e. attributes that contain sensitive information about an individual, e.g., diseases, political or religious views, income.

In presence of identifiers the re-identification risk is clearly maximum (i.e., $P = 1$), but even if identifiers are removed, the combination of QIs can lead to the identification of individuals and this implies a high risk. The k-anonymity condition requires that *every* combination of QIs is shared by at least k records in the dataset. A large k value indicates that the dataset has a low re-identification risk because an attacker has a probability $P = 1/k$ to re-identify a data entry (i.e., associate the sensitive attribute of a record to the identity of a User). Therefore, the (re-identification) risk related to a k-anonymous data-view v is:

$$risk(v) = 1/k_v \times I$$

where I is the impact associated with the identification of the users in the dataset. The severity of the impact is often evaluated in terms of monetary cost but it can also be assessed by assigning severity levels (e.g., spanning from minimal to critical). In this paper we will evaluate the impact in the interval $[0, 1]$, where 0 is *minimal* impact and 1 is *maximal* impacts. For the sake of simplicity, we will set the impact $I = 1$ and this will allow us to normalize the risk and the trust values to $[0, 1]$. For a discussion on the impact of normalization, see [3].

---

[2] Other privacy metrics exist (for example, ℓ-diversity, and t-closeness, see [19] for a survey), but k-anonymity is still a *de-facto* standard in real-world applications.

Trust Model. In our framework trust plays the role of a risk threshold: trusted users are allowed to take large risks.[3] We assign a trust level $T_{user}(u)$ to user u depending on her competence/roles and the tasks these roles are expected to fulfill (see Table 1). Following the data minimization policy, a role should have *enough* trust to access the resources (data) needed to fulfill these tasks and not more. These values are assigned on a scale from 0 to 1, where 0 means that no privacy risk can be taken and 1 means the user can be granted access to a maximum amount of data.

Notice that the same request can be used to fulfill different tasks in different contexts. To illustrate, consider the tasks *"Perform Maintenance and Improvement tasks"* and *"React to a Security Incident"*. In the latter, the need to react to a security threat overcomes the privacy requirements. The request could be granted access to sensitive data and therefore can be given a higher level of trust. We will define the two context-related trust levels as $T_{context}(Alert) = 1$ and $T_{context}(noAlert) = 0$

To compute the request trustworthiness (*total trust value*) we can use the approach for multi-dimensions trust computation proposed in [26], where the total trust is computed as a weighted sum of trust factor values.

$$T = \sum_{i=1}^{n} W_i \times T_i(\beta_i)$$

where β_i, $T_i()$, and W_i is a trust factor, a trust function and the weight of the i-th trust factor for $i = 1, \ldots, n$ respectively, subject to the constraint $\sum_{i=1}^{n} W_i = 1$. In our case, $n = 2$ and we can express our total trust value as:

$$T(q) = W \times T_{user}(u) + (1 - W) \times T_{context}(c)$$

Adjustment Strategies

Risk Mitigation. A possible way to decrease the disclosure risk is anonymization. Anonymization is a commonly used practice to reduce privacy risk, consisting in obfuscating, in part or completely, the personally identifiable information in a dataset. Anonymization methods include [13]:

- *Suppression:* Removal of certain records or part of these records (columns, tuples, etc., such UserId column);
- *Generalization:* Recoding data into broader classes (e.g., releasing only a Network prefixes instead of IP addresses etc.) or by rounding/clustering numerical data;

Traditionally, anonymization is run off-line, but more recently risk-based access control models, which use in-the-fly anonymization as mitigation strategy have been proposed [4].

---

[3] For a discussion on how relating this definition with more classical trust metrics, see [3]. See [21] for a survey of different approaches to defining trust.

Trust Enhancement. Enhancing the trust results in raising the risk threshold and adopting a more permissive evaluation. In return, proofs and/or guarantees limiting possible misuse scenarios must be provided; for instance by asking the user to provide a stronger authentication to limit the likelihood of an identity theft and temporarily increase the trust of a user T_u, which impacts the trust value according to Eq. 3.2. We can also provide restricted access to a resource for a determined amount of time, then delete the resource (data), this represents a change in the context and, accordingly, it increases T_C impacting the request trust value as well. Trust enhancement mechanisms can be implemented in the form of access and usage control obligations.

In the standard XACML [17], access control obligations are defined as parts of policies and included in authorization responses created by the PDP; they are enforced by the PEP on behalf of the subject issuing the authorization request. Besides, their application as an outcome of the authorization decisions, obligations may also be applied during or after the consumption of a requested resource or the execution of a requested operation [1,30]: for example, a policy may state a specific retention period for any copy of a resource whose access was granted to the requester. In these cases, a trusted component must exist that could operate in real time as a PEP. This situation is generally referred as *Usage Control* (UC) [37].

4   Framework Architecture

In this section, we present an abstract architecture for our risk-based privacy-aware access control framework, and we explain the different steps of the data request evaluation work-flow.

Figure 2 depicts the four main modules of our framework:

Risk-Based Access Control Module. This module is the entry point of the framework. It intercepts each data request to perform the access evaluation. Access will be fully grant, partial/conditionally grant, or denied (see Eq. 1) following a risk-based approach were the request risk and trustworthiness are assessed and compared, possibly after applying the adequate adjustment strategy (set of operations aiming to lower risk and/or enhance trust).

This module is composed of three main components inspired by the XACML standard reference architecture[4] (PEP, PEP, and PIP), which we adapted to the risk-based authorization model that requires more complex operations, such as risk and trust assessment and adjustment. We call the modified components respectively RBA-EP (Risk-Based Authorization - Evaluation

---

[4] In the XACML3.0 (eXtensible Access Control Markup Language) standard [17] the PDP is the point that evaluates an access request against an authorization policy and issues an access decision and the PEP Policy Enforcement Point is the point that intercepts user's request call the PDP for an access decision then enforce this decision by allowing or denying the access. The PIP is the point that can be called to provide additional information about the resource, requester or environment.

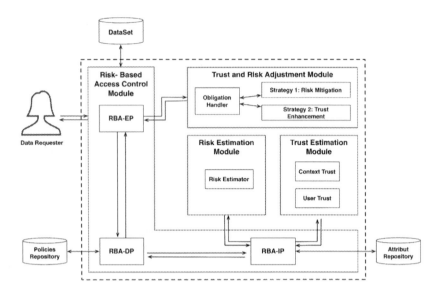

Fig. 2. Architecture of the risk-based privacy-aware access control framework.

Point), RBA-DP (Risk-Based Authorization - Decision Point), and (Risk-Based Authorization - Information Point).

Risk Estimation Module. This module is used to assess the level of risk, based on the data requested, context and criteria defined in the risk estimator configuration. For example, for re-identification risk, this configuration should include the metrics to be used (e.g., k-anonymity, ℓ-diversity) and a domain-specific classification of data attributes indicating which attribute is to be considered as *sensitive*, *QIs* and so on. To estimate risk, this module can require additional information about requester and context from the RBA-IP.

Trust Estimation Module. This module is used to assess the trust level of a request. In particular based on user attributes like role, and past behavior. Trust computation can also take into account context attributes, for instance, in our case, access context (purposes) e.g., access is requested for maintenance of a pattern, and security context e.g., access is requested during a security alert.

Trust and Risk Adjustment Module. This module is activated by the Risk-Based Access Control module (more precisely by the RBA-EP) to adjust risk and/or trust levels, when the access risk to the requested resource exceeds the trust level, in such a case, two possible options are available:

– *Decrease Risk:* if this option is selected this module, first, produces an estimation of the minimal anonymization level to be applied in order to meet the required risk level. (i.e., in case of k-anonymity metric, the risk estimation module computes the minimal value of k that respects the risk threshold constraint $R \leq T$ see Eq. 1). Then, the optimal risk mitigation

operations are applied (e.g., anonymization operations, which decrease risk but minimize the information loss).
– *Enhance Trust:* the trustworthiness estimation can be increased in return of the execution of certain operations. Before granting access to the resource (e.g., second-factor authentication) at the moment of the access (e.g., monitoring or notifications) or when specific events occur after granting access; for example, in usage control, we may prescribe the deletion of a resource after that a retention period expires.

In Fig. 3, we illustrate the interactions between different modules of the framework during the request evaluation and decision enforcement: First, the Risk-Based Access Control module, more precisely the RBA-EP, extracts the request target (i.e., resource, subject/requester and environment/context attributes) and sends the information to the RBA-DP for evaluation. The RBA-DP checks if there are any access policies/rules matching this target in the Policy Repository. During the matching, the RBA-IP can be called to resolve or provide more information about some attribute. If the matching fails (e.g., missing or unknown attributes) or if the target matches a policy denying the access to the specific target (Blacklisted Target), a *deny* response is sent to the requester. If a match is found and the policy does not explicitly deny access, we need to compare the risk and trust levels to check if we should *grant* access or if we need to apply adjustments to these levels (see Sect. 3.1 (Eq. 1)). To this aim, the Risk-Based Access Control module calls the Risk Estimation Module to determine the risk level of the request and the Trust Estimation Module to determine the requester and context trust. Then, Trust and Risk Mitigation Module enters into play to increase trust and/or reduce risk, if necessary, before granting partial or conditional access to the resource.

5 Policy Implementation

In this section, we present a possible way of expressing the authorizations model, described in Sect. 3 in Eq. 1, through risk-based policies.

Since we based the access control model and architecture on a modified version of XACML's architecture, we will also propose an extended version of XACML's language to implement our policies. We propose to organize policies according to *patterns*, i.e., each policy expresses the authorizations required to run and investigate a pattern. This choice allows for a flexible policy management (addition, modification, and deletion), in fact, new patterns are often added and old once updated or removed from the system and each of these operations requires an update of the policies. Hence if we dedicate a policy for each pattern, when an existing pattern is modified or a new one is created, we just need to revise the policy expressing the authorizations required by the pattern or create a new policy.

The proposed policies have a similar structure as described in Example 1.1. In this policy example, we express the authorizations required by the pattern detecting *Brute Force Attacks*.

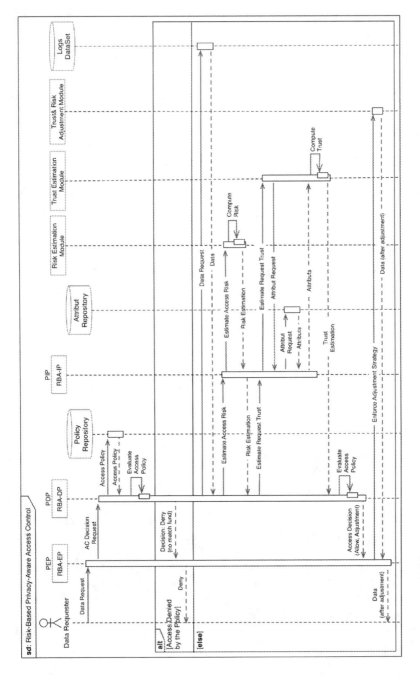

Fig. 3. Sequence of interactions in the risk-based privacy-aware access control framework.

Example 1.1. Risk-Based Policy Sample: Brute Force Attack Pattern

```
1   <!-- Brut_force_attack policy -->
2   <Policy PolicyId="brut_force_attack_policy" RuleCombiningAlgId="
        permit-overrides">
3       <Target> ... </Target>
4       <Rule RuleId="deny_all" Effect="Deny">
5       <!-- Deny access for the policy Target -->
6       ...
7       <Rule RuleId="allow_trust_higher_than_risk" Effect="Permit">
8       <!-- Allow access if trust>=risk -->
9       ... </Rule>
10      <Rule RuleId="adjust_risk_higher_than_trust" Effect="Permit">
11      <!-- Adjust Trust or Risk Values if trust<risk -->
12      ... </Rule>
13      <Rule RuleId="exceptional_rule_1" Effect="Permit">
14      <!-- optional -->
15      <Target> <!-- exception's target --> </Target>
16  </Rule>
17  </Policy>
```

Each policy is composed by a main **target** and three **rules** expressing the three possible outcomes of the access evaluation, i.e., *deny*, *grant*, or apply an adjustment strategy σ (see Sect. 3.1). Rules will apply to the same target defined as the policy target. We use the combination algorithm *permit-overrides* to select the rule to enforce, in case more then one rule is applicable (i.e., rule's target matches the request's target and the rule's conditions are satisfied by the request). The algorithm *permit-overrides* enforces the first rule that permits access (allows access) once the obligations defined by the rule are enforced (see [17] for more details). If no rules permit access that the first rule denying access will be enforced. Some patterns may require additional *exceptional* rules (i.e., break the glass rules) usually more permissive and with a more specific target. For instance, for very critical patterns (e.g., a denial of service attack) we can allow a super administrator to access the data for investigation without trust assessment despite the risk level (see Example 1.6).

The main target describes which subjects (requesters), which resources and what actions the policy applies to. In example (Example 1.1), for instance, the policy applies to any requester with any role known to the risk-based access control system *ROLE:ANY* (see Example 1.2). The targeted resource is the table containing the logs *TABLE:LOGS* and the targeted action is *ACTION:READ* access. The policy target should also specify the context of the access which can express for instance the access purpose e.g., *running* or *investigating* the pattern *PATTERN:BF-ATTACK*. It could be also used to describe the security context e.g., *alert* or *logged event*. The policy target (or main target) can be refined within the rules. For instance, the exceptional rule target (Example 1.6) refers to the group of subject with the role of *ROLE:SUPER_ADMIN* which is more specific than *ROLE:ANY*, it also narrows the context and make the rule only applicable in case of a security alert *SEC_CONTEXT:ALERT*.

Example 1.2. Policy main target

```
1   <Target>
2       <AnyOf> <AllOf> <Match MatchId="string-equal">
3         <AttributeValue DataType="string">ROLE:ANY</AttributeValue>
4         <AttributeDesignator AttributeId="subject-role" Category="
                subject" .../>
5       </Match> </AllOf> </AnyOf>
6       <AnyOf> <AllOf> <Match MatchId="string-equal">
7         <AttributeValue DataType="string">TABLE:LOGS</
                AttributeValue>
8         <AttributeDesignator AttributeId="resource-id" Category="
                resource" .../>
9       </Match> </AllOf> </AnyOf>
10      <AnyOf> <AllOf> <Match MatchId="string-equal">
11        <AttributeValue DataType="string">ACTION:READ</
                AttributeValue>
12        <AttributeDesignator AttributeId="action-id" Category="
                action" .../>
13      </Match> </AllOf> </AnyOf>
14      <AnyOf> <AllOf> <Match MatchId="string-equal">
15        <AttributeValue DataType="string">PATTERN:BF-ATTACK</
                AttributeValue>
16        <AttributeDesignator AttributeId="pattern-id" Category="
                environment" .../>
17      </Match> </AllOf> </AnyOf>
18  </Target>
```

The first rule, *Deny Rule* in the Policy (Example 1.3) does not specify the target, hence, inherits the policy's target. This rule does not have any conditions or obligation its aim is to guarantee that the access is denied if none of the other rules allow it (e.g., the conditions of other rules were not satisfied or errors occurred during the rules evaluation).

Example 1.3. Deny Rule

```
1   <Rule RuleId="deny_all" Effect="Deny">
2       <!-- Deny any access to all Roles/Subjects in the Target -->
3   </Rule>
```

The *Adjust Rule* (in Example 1.4) applies to the policies target as well. It expresses the second outcome in Eq. 1, where an adjustment strategy σ is required to be applied before granting access. The condition of application of this rule (in Line 3), is that the trust is lower than the risk level. To check this condition, we need, first, to compute the trust and risk values, which as indicated in Sect. 4, is the task of the *RBA-IP*, however we need to indicate to the *RBA-IP* where to possibly find the information, e.g., we define that the *request-trust* can be computed using the *TrustAssessmentModule* (Line 8) and the *request-risk* can be computed by the *RiskAssessmentModule* (Line 15). In case of failure to compute the trust level, minimum trust level $T = 0$ will be assigned to the request, and if the failure occurs in the risk computation, we will assign to the request maximum risk level $R = 1$. If the condition of *Adjust Rule* is fulfilled then access cannot be granted to the resource, unless an adjustment phase is successfully carried. The adjustment strategies for each pattern are expressed through obligations (see Example 1.7, 1.8, or 1.9).

Example 1.4. Adjust Rule

```
1  <Rule RuleId="adjust_trust_lower_than_risk" Effect="Permit">
2  <!-- Ajust Trust or Risk Values if trust < risk -->
3      <Condition><!-- applicable if trust is lower than risk -->
4          <Apply FunctionId="double-greater-than-or-equal">
5              <Apply FunctionId="function:or"> <!-- compute the trust
                   or trust =0 -->
6                  <!-- call TrustAssessmentModule to compute the
                       trust -->
7                  <Apply FunctionId="function:double-one-and-only">
8                      <AttributeDesignator Category="request-trust"
                           AttributeId="trust" Issuer="
                           TrustAssessmentModule"/>
9                  </Apply>
10                 <AttributeValue>0</AttributeValue>
11             </Apply>
12             <Apply FunctionId="function:or"> <!-- compute risk or
                   risk =1 -->
13                 <!-- call RiskAssessmentModule to compute risk -->
14                 <Apply FunctionId="function:double-one-and-only">
15                     <AttributeDesignator Category="request-risk"
                           AttributeId="risk" Issuer="
                           RiskAssessmentModule"/>
16                 </Apply>
17                 <AttributeValue>1</AttributeValue>
18             </Apply>
19         </Apply>
20     </Condition>
21     <ObligationExpressions>
22         <!--Adjustment Strategies-->
23     </ObligationExpressions>
24 </Rule>
```

The *Allow Rule* (see Example 1.5) expresses the last out come of evaluation in the authorization model (Eq. 1). Similarly to the *Deny Rule* and *Adjust Rule*, this third rule, has the same target as the policy. According to *Allow Rule*, access to the requested data is fully granted if the trustworthiness level of the request is higher than its risk level. The trust and risk levels assessment is expressed the same way as the *Adjust Rule* in Example 1.4 (Lines 4 to 18).

Example 1.5. Allow Rule

```
1  <Rule RuleId="allow_trust_higher_than_or_equals_risk" Effect="
       Permit">
2  <!-- Allow access if trust >= risk -->
3      <Condition><!-- applicable if trust is higher than or equals
           risk -->
4          <Apply FunctionId="double-greater-than-or-equal">
5              <!-- compute and compare trust and risk levels -->
6          </Apply>
7      </Condition>
8  </Rule>
```

Obligations, in the XACML standard, are enforced by the PEP immediately *after* granting or denying access, e.g., allowing access to a user *Alice* with the obligation to log *Alice*'s actions during the access session. However, our authorization model needs, in some cases to enforce certain actions *before* granting access, such as transformations on data, and other actions during the consumption of the data. Thus we propose to use two other types *pre-decision* and *post-decision* obligations categories. These new obligation categories were inspired by [44]. In Table 2 we provide a description for each category, we also provide an implementation example in Examples 1.7, 1.8 and 1.9.

We discussed in Sect. 6.5 possible ways to dynamically select the mitigation strategies based on utility, but we are not including the implementation in this paper.

Table 2. Obligation types

At-decision obligations	Are similar to the classic XACML3.0 obligations they are actions to be enforced at the same time then the access decision e.g., sending notifications, logging session details. These obligations fulfillment do not influence the access decision
Pre-decision obligations	Are actions to be enforced before enforcing the access decision to a resource e.g., anonymization, encryption, requesting a stronger authentication. The success or failure to fulfill these obligations can influence the access decision
Post-decision obligations	Are actions expected to be enforced after enforcing the access decision e.g., deletion of the data

Example 1.6. Exceptional rule allowing the super admin to access without risk and trust assessment

```
1   <Rule RuleId="exceptional_rule_1" Effect="Permit">
2   <Target>
3       <AnyOf> <AllOf> <Match MatchId="string-equal">
4           <AttributeValue DataType="string">ROLE:SEC_ADMIN</
            AttributeValue>
5           <AttributeDesignator AttributeId="subject-role" Category="
            subject" .../>
6       <AnyOf> <AllOf> <Match MatchId="string-equal">
7           <AttributeValue DataType="string">SEC_CONTEXT:ALERT</
            AttributeValue>
8           <AttributeDesignator AttributeId="pattern-id" Category="
            environment" .../>
9       </Match> </AllOf> </AnyOf>
10  </Target>
11  </Rule>
```

Example 1.7. at-access Obligations

```
1   <ObligationExpression ObligationId="system:log" ObligationType="at-
        access" FulfillOn=Permit>
2       <!-- Temporairly Grant higher trust level -->
3       <!-- Log the access request and access session -->
4   </ObligationExpression>
```

Example 1.8. pre-access Obligations

```
1  <ObligationExpression ObligationId="system:anonymize" ObligationType="
       pre-access" FulfillOn=Permit>
2      <AttributeAssignmentExpression><!--compute required anonymity level
           -->
3          <AttributeDesignator AttributeId="optimal_k" Issuer="
               TrustAndRiskAjustementModule" />
4      </AttributeAssignmentExpression>
5      <AttributeAssignmentExpression><!-- apply anonymization -->
6          <AttributeDesignator AttributeId="anonymizer:k-anonymity" Issuer
               ="TrustAndRiskAjustementModule" />
7      </AttributeAssignmentExpression>
8  </ObligationExpression>
```

Example 1.9. post-access Obligations

```
1  <ObligationExpression ObligationId="remote-rba-ep:data-deletion"
       ObligationType="post-access" FulfillOn=Permit>
2      <AttributeAssignmentExpression><!-- Enhance Trust-->
3          <AttributeDesignator AttributeId="enhanced-trust-level"
               Issuer="TrustAndRiskAjustementModule"/>
4      </AttributeAssignmentExpression>
5      <AttributeAssignmentExpression><!-- fix access time window-->
6          <AttributeDesignator AttributeId="time-window" Issuer="
               TrustAndRiskAjustementModule"/>
7      </AttributeAssignmentExpression>
8      <AttributeAssignmentExpression><!-- delete data-->
9          <AttributeDesignator AttributeId="action:data-deletion"/>
10     </AttributeAssignmentExpression>
11 </ObligationExpression>
```

6   Experimental Evaluation

We validate our approach by applying the model described in Sect. 3 to the scenario described in Sect. 2. The TDS is expected to provide accurate real-time results, therefore we investigate the impact of our approach on the functioning of the TDS, in particular, whether the expected *Performance* and *Utility* matches the accuracy and real-time requirements.

More in details, as mentioned in Sect. 2, the TDS allows to detect potential attack patterns automatically, and then if additional investigations are needed, a human operator can browse the log data of the events corresponding to a given pattern for manual inspection.

Ideally, the operator should be able to perform the manual investigation (i.e., decide if the detected threat is a false or true positive). Some investigations can be conducted on data where the personal information are anonymized (or in any case, where the re-identification risk is low). If the operator does not have sufficient information to decide, he/she should be granted access to less anonymized (riskier) data, or in other words get higher access privileges (trust enhancement) acquiring administrator rights, or directly involving an administrator.

Accordingly, we need to check:

– *Utility*. Does the model allow a low trusted operator (i.e., small risk threshold) to perform the investigation in most cases, and relying on trust enhancement for the remaining cases?

– *Performance.* Does the additional anonymization step impact real-time performance?

Before addressing these questions (see Sect. 6.6), we need to describe our prototype implementation (Sect. 6.1), the data set and its attributes classification from a privacy risk perspective (Sect. 6.2), the selection of typical patterns used for the validation (Sect. 6.3), the utility measure (Sect. 6.5) and the trust level setting (Sect. 6.4).

6.1   Prototype Implementation

We developed a prototype of our framework, based on the implementation described in [5]. Our prototype is implemented in Java 8 and uses SAP HANA Database. It is composed of 3 main modules:

– The *Risk Aware Access Control module:* mimics a typical XACML data flow, providing an implementation of the PDP, the PEP, and the PIP functionality as well as a set of authorization policies.
– The *Risk Estimation module:* evaluates the privacy risk using pre-configured criteria (privacy metrics, anonymization technique, identifying information). It compares the privacy risk to the request trustworthiness level, then produces an estimation of the minimal anonymization to be applied in order to meet this level.
– The *Trust & Risk Adjustment module:* we implemented the Risk Adjustment Component to perform anonymization. It uses ARX [23] a Java anonymization framework implementing well-established privacy anonymization algorithms and privacy metrics such as k-anonymity, ℓ-diversity, t-closeness, etc. (the Trust Adjustment Component was not implemented in this version of the prototype.)

6.2   Data Set and Privacy Classification

To test the performance of our framework in the TDS use case, we used a dataset containing around $1bn$ record of log data collected from SAP systems deployed in a test environment[5]. The logs dataset is composed 20 fields (in Table 3 we present a summary of the most important fields)

As described in Sect. 3.2, to anonymize a dataset, we first need to formalize our assumptions on the attributes that can be used to re-identify the entry, or, in other words, classify the attributes in terms of identifiers, QIs, and sensitive attributes. This classification, typically, depends on the specific domain. QIs should include the attributes a possible attacker is likely to have access to from other sources, whereas sensitive attributes depend on the application the anonymized data are used for. For example, in our experiments, we set (obviously) User ID as an identifier and the IP address as a quasi-identifier. Similarly, we assume that the Transaction name (the called function) cannot provide any

---

[5] For an analysis of the performance of the model on a benchmark dataset see [4].

Table 3. An extract of the Log dataset columns, privacy classification of each column and anonymization technique to be applied

Log Events data set		
Attribute	Type	Anonymization
EventID	Non-Sensitive	
Timestamp	Sensitive	
UserId (Origin)	Identifier	Suppression
UserId (Target)	Identifier	Suppression
SystemId (Origin)	QI	Generalization
SystemId (Target)	QI	Generalization
Hostname (Origin)	QI	Generalization
IPAddress (Origin)	QI	Truncation
MACAddress (Origin)	QI	Truncation
TransactionName	Sensitive	
TargetResource	Sensitive	

help for re-identification, therefore we consider it a sensitive attribute (and no anonymization will be applied). Table 3 provides an example of this classification, and, for identifiers and quasi-identifiers, the corresponding anonymization methods applied.

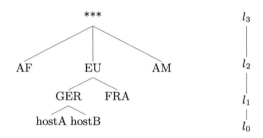

Fig. 4. The generalization hierarchy for host names is organized as following: l_1 and l_2 are a location based generalization by country then by continent. in level l_3 host names are totally obfuscated and entirely revealed at the level l_0.

6.3   Pattern Detection and Investigation

In our experiments, we focus on 5 typical *Patterns* with different complexity in terms of the size of the returned data-views and the privacy risk. Two different kinds of queries are used during each phase respectively *Detection Queries* and *Investigation Queries*. The selected queries {**Q1** ... **Q5**} described in Table 4 are all *Investigation Queries*. An *Investigation Query* is a "SELECT *" extracting all the details of the events corresponding to certain pattern.

Table 4. Queries: resulting views size and risk level

Query	Corresponding pattern	View size	Risk level
Q1	Brute force attack	Large (50550)	Very high ($k = 2$)
Q2	Security configuration changed	Large (40300)	Medium ($k = 7$)
Q3	Blacklisted function called	Medium (14500)	Very high ($k = 1$)
Q4	Table dropped or altered	Small (228)	Medium ($k = 6$)
Q5	User assigned to admin group	Very small (12)	Very high ($k = 1$)

6.4   Roles and Trustworthiness Levels

We have 3 roles Operator, Administrator and an Advanced administrator with increasing access requirements (to fulfill their tasks), therefore increasing privacy clearances, (i.e., larger risk tolerance). Usually, for k-anonymity, k values in the range 3–10 are considered medium risk, $k > 10$ low risk, and for $k \leq 2$ the risk is very high (clearly, for $k = 1$ the risk is the maximum, no anonymity) [33]. Therefore we propose the parameter setting described in Table 5, where for sake of simplicity we have considered a single trust factor $T = T_u$ (i.e. we set $W = 1$ in Eq. 3.2).

Table 5. Users/roles privacy clearances and trustworthiness levels

Role	Access requirement	Privacy clearance	Trust level (Risk threshold)
Operator	Low	Minimal ($k > 10$)	$T_u \in [0.05, 0.1]$
Administrator	Medium	Medium ($k > 2$)	$T_u \in [0.1, 0.5]$
Advanced administrator	High	Maximum ($k \leq 2$)	$T_u \in [0.5, 1]$

6.5   Utility Evaluation

The effect of anonymization in terms of utility is a widely discussed issue in the literature several generic metrics have been proposed to quantify the *"damage"* caused by anonymization (see [20] for a review). However, these metrics do not make any assumption on the usage of the data (so-called *syntactic metrics*), limiting their applicability on realistic use-cases.

Other approaches propose to assess the accuracy loss (Utility loss) of a system (i.e., IDS in [25], Classifier in [9]) by comparing the results of certain operations run on original then anonymized dataset using use case related criteria (i.e., in the context of a TDS the comparison criteria can be the number of *False positives*). Although interesting for our context, this approach cannot be applied in our use case, since it assumes that the analysis is run directly on anonymized data, whereas, in our use case, the pattern detection is performed on *clear* data, and the anonymization is applied only on the results (data-view).

We propose a method combining both approaches and that would include an evaluation:

- *From Syntactic standpoint:* The information loss caused by the anonymization, we use the precision metric that allows us to estimate the precision degradation of QIs based on the level of generalization with respect to the generalization tree depth (e.g., for the generalization tree Fig. 4 if we allow access to continent instead of host-names we used the 3rd level generalization out of 4 possible levels so $d_p(hostnames) = 3/4 = 75\%$ precision degradation for host-names).
- *From Functional standpoint:* The effect of this loss on our use case. During the investigation phase, the operator, mostly, bases their analysis on a subset of attributes, which are different for each attack pattern. Thus we will assign a utility coefficient uc to different attributes based on the relevance of the attribute to the pattern/query.

Combining the two approaches we compute the utility degradation of a data-view v as

$$U_d(v) = \sum_{a_i \in A} uc_{a_i} \times d_p(a_i) \tag{2}$$

with $A = \{a_1..a_i\}$ the set of attributes in the data set. We also set the precision degradation of the identifiers to $d_p(identifiers) = 1$ as they will be totally suppressed after the anonymization.

6.6 Results and Analysis

For our experiments, we want to investigate: *(i)* Performance: the impact of on-the-fly anonymization (as risk mitigation strategy) on the performance (response time). *(ii)* Utility: we would like to investigate if the quality of resulting data is generally enough to fulfill the expected tasks for every user/role for various pattern investigation.

In order to evaluate these aspects we run several experiments considering 5 patterns and 7 users/role with different trustworthiness level, $t = \{0.055, 0.083\}$ Operators, $t = \{0.12, 0.15, 0.45\}$ Administrators, and $t = \{0.9, 1\}$ Advanced Administrators. The corresponding size and anonymity level of the views returned by the queries (corresponding to the selected patterns) are reported in Table 4. In the rest of this section we will indicate both the queries and the corresponding views as **Q1**, **Q2**, **Q3**, **Q4** and **Q5**.

Performance and Scalability: To evaluate the performance of our tool, including the computational overhead caused by the anonymization, we run queries **Q1**, **Q2**, **Q4**, and **Q5** (described in Table 4) using our access control prototype experiment, 100 times for each query to average out the variance of

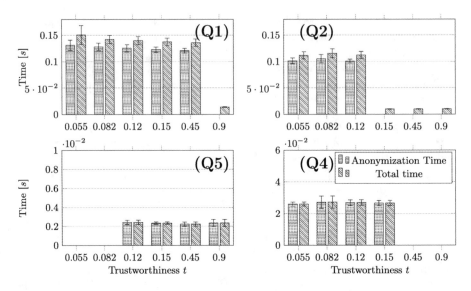

Fig. 5. Average anonymisation time (horizontal striped bars) and average total response time (diagonally striped bars) for **Q1**, **Q2**, **Q4**, and **Q5** (data-views) and 6 different users (trust levels).

the response time. In Fig. 5 we report the results of the experiments for the four queries for the 6 trustworthiness levels.

For **Q1**, we observe that the anonymization process increases significantly the response time. In fact when the query is carried out by the most trusted user ($t = 0.9$), with no anonymization needed, the response time on average is less than 15 ms (see Fig. 5**Q1**, the diagonally striped bar corresponding to $t = 0.9$). By decreasing the trustworthiness of the requester the view must be anonymized and the average response time increases to 150 ms in the worst case (cf. Fig. 5**Q1**, the diagonally striped bar corresponding to $t = 0.055$). This time difference is entirely due to the anonymization time (130 ms, as shown in Fig. 5, **Q1**, horizontal striped bars corresponding to $t = 0.055$). Increasing the trust level decreases the needed anonymization, but it slightly affects anonymization time. We can observe a similar behavior in the other queries (see Fig. 5**Q2**, **Q4**, and **Q5**), with an increase of response time when anonymization takes place and no significant variations in performance for different levels of anonymization. For instance, for **Q2** and **Q4** we have two views with an already medium level of anonymity (respectively $k = 7$ and $k = 6$), the anonymization (when needed) still impacts the performance in the same scale then **Q1** and **Q5** with very low anonymity level (respectively $k = 2$ and $k = 1$).

From these experiments, we observe that when anonymization is applied the response time increases, but, even in the worst cases, the increase is far less than one order of magnitude, and, basically, it has no impact on the real-time response of the system. Moreover, the application of different levels of anonymization

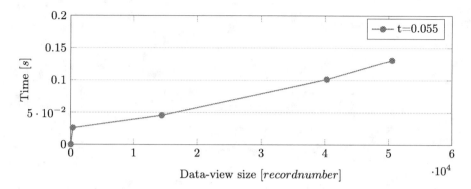

Fig. 6. Average anonymization time variation according to data-view sizes (for trust-worthiness $t = 0.055$).

(different k in our case) has a small impact. We will investigate in the next paragraph the effect of the data-view size on the Anonymization and Response time.

Let us analyze the behavior of the anonymization time increasing the size of the dataset. Typically patterns run in the limited time window (e.g., 10 to 30 min) producing small-sized data-views (i.e., in the range of $10 - 10^3$). To investigate the scalability of our approach, in Fig. 6, we report the average anonymization time variation for 5 different data-view {**Q1** to **Q5**} (with 5 different sizes see Table 4) and a low trustworthiness level ($t = 0.055$, so anonymization is always applied). As mentioned above, the worst case (around $5 \cdot 10^4$ records) takes less than 150 ms, and a linear extrapolation of the data allows as to estimate the anonymization time for a 10^5 data view (so, 100 times the typical size) around 200 ms, which it can be safely considered as a real-time response for our use case.

Utility: Trustworthiness levels (i.e., risk threshold) should be set to allow the best a trade-off between data exploitation and privacy protection. In our use case we set our trustworthiness levels respecting a conventional distribution of privacy risk levels presented in Table 5, and we would like to investigate the convenience of this repartition by answering the following question: Do these trustworthiness levels provide enough data (or data with enough utility) to allow each user/role to fulfill their tasks described in Table 1. In Fig. 7, we report the utility degradation according to the six selected trustworthiness levels, representing the 3 roles (reported on the top of the figure). We can observe that the utility degradation (obviously) decreases as we increase the trust level, with the limiting case of $t = 1$ with no utility loss (and no anonymization) for the Advanced Administrator. For most of the patterns (4 over 5, so except **Q5**), the Operator role has a maximum utility loss of 30%, showing that the specific anonymization transformations applied are strongly decreasing the risk, and limiting the impact on

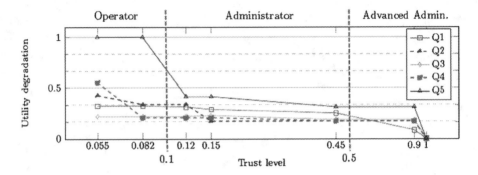

Fig. 7. Utility degradation by trust level for different queries

the utility. That should allow performing the analysis on the anonymized data, without the need to enhance the trust level (so no need to get Admin rights).

In the case of **Q5**, the anonymization is not able to significantly decreases the risk, without largely impacting the utility. In fact, the Operator is left with no information (utility degradation = 1), and to analyze the result an increase of the acceptable risk threshold (trust level) is needed. Enhancing trust (i.e. assigning Admin rights to the Operator) could reduce the utility degradation in the 30%–40% range, likely allowing the assessment of the pattern result. We should note, that **Q5** is particularly hard to anonymize, because it has fewer events (around 10), and, since k-anonymity is a measure of indistinguishability, it needs strong anonymization.

Figure 7 also shows that in most cases increasing the trust level for Administrator or even Advanced Administrator (except of course for $t = 1$, where we have no anonymization) the impact on utility degradation is moderate: for example **Q1** and **Q4** are almost flat in the Administrator zone, similarly **Q2** has a first drop, and stays flat in the Administrator and Advanced Administrator parts. In other words, increasing the risk thresholds, we could take more risk, but we do not gain much in terms of the utility. This counter-intuitive effect is mostly due to the difficulty to find an anonymization strategy able to equalize the risk threshold. As mentioned in Sect. 3.1, in practical cases the number of possible anonymization strategies is limited, and to fulfill the condition of Eq. 1 the final risk may be quite below the risk thresholds (trust values). In practice, in many cases, even increasing the risk thresholds (trust values), it is not possible to find a more optimal (from the utility point of view) anonymization strategy. In Fig. 7 we show the utility loss for four patterns both showing the risk thresholds (dotted lines) and the *actual* risk achieved after the anonymization. In the ideal case, the two curves should be the same, meaning that we could always find a transformation that equalizes actual risk and risk thresholds (trust), but in practice, we see that we are often far from this optimal condition. For example, for pattern **Q2**, with risk thresholds $t = 0.15$, $t = 0.45$ (Administrator role) and $t = 0.9$ (Advanced Administrator), indicated with red circles, we have the same value of utility degradation. In fact, the anonymization strategy found for $t = 0.15$

case, corresponds to an actual risk of 0.14 (square dots with a circle in Fig. 7, upper-right panel), so quite close to the threshold. Increasing the thresholds to $t = 0.45$ and $t = 0.9$ (round dots with a circle in the figure), no better strategies were found, so the same anonymization strategy is applied, and clearly, the final risk is still 0.14 (and utility is the same), well below the thresholds. Similar effects are also present in the other patterns.

The experimental analysis shows that adapting the anonymization to the specific patterns, we can mostly preserve enough information for the investigation, keeping the privacy risk low. In cases where this is not sufficient, typically characterized by small data set, the trust enhancement strategy can support the access to less-anonymized data.

7   Related Work

Privacy Issues in Intrusion Detection: Privacy issues related to sharing and/or using network and log data in IDSs and TDSs have received a growing interest in the last few years. Several models were proposed in the literature to describe privacy breaches related to the sharing and using of log data and privacy-preserving approaches have been proposed to address these issues.

A strict enforcement of the *need-to-know* principle has been proposed for reducing the likelihood of privacy violations. For example, Ulltveit-Moe et *al.* in [42] propose to set two profiles of users according to the expertise level: the first profile allows monitoring tasks using anonymized data the second consists of security experts, with clearance to perform necessary privacy-sensitive operations to investigate attacks. This model clearly increases the privacy protection, but it is hard to apply in realistic cases, since it relies on anonymizing the entire (source) dataset beforehand, resulting in either low privacy or low utility. In our approach, we use a similar approach, strictly adopting the need-to-know principle, but, as described in Sect. 3, the anonymization is dynamically only in the dataset resulting from a pattern, and according to the *trust* level of the users/roles. As a result, we can use the *better* anonymization transformation depending on the specific utility of each pattern, assuring an increase of both privacy and utility.

Other works focus on specific anonymization techniques for logs (see [29] for review), and on measuring the privacy risk. For example, in [41], the authors use entropy to measure privacy leakage in IDS alerts. We implemented several of the proposed anonymization techniques in our prototype, and, although based on k-anonymity, our framework can include other privacy measures by changing the risk function. More specifically, entropy-based privacy metrics can be easily integrated with k-anonymity approach, as shown in [24].

Risk Based Access Control Systems: Several risk and trust-based access control models have been introduced in the last years. (e.g. [10–12,16,39]), where for each access request or permission activation, the corresponding risk is estimated and if the risk is less than a threshold (often related to trust) then the operation is permitted, otherwise, it is denied. Cheng et al. [12] estimate risk and trust thresholds from the sensitivity labels of the resource and clearance level of the users in a multi-level-security system. Cheng's model allows users to spend *tokens* to access resources with a risk higher than their trust level. To mitigate this risk the authors propose three categories of actions meant to deter the users from disclosing the data after granting access to the data. The details on how these actions are implemented and applied in real cases are not provided. In our approach, we propose concrete actions both deterring actions (e.g., monitoring) to enhance trust but also preventive actions (e.g., anonymization) to lower the risk.

Chen et al. [10] introduced an abstract model which allows role activation based on a risk evaluation compared to predefined risk thresholds. Trust values are included in the risk calculation. If the risk is too high, the model includes mitigation strategies, indicated as system obligations. The paper does not specify how to compute the risk thresholds, trust, nor the structure of obligations. In a derived model [11], mitigation strategies have been explicitly defined in terms of user obligations (actions that have to be fulfilled by the user). The model also introduces the concept of *diligence score*, which measured the diligence of the user to fulfill the obligations (as a behavioral trust model), and impact the risk estimation. In addition to this type of obligations (post-decision obligation), we propose to use two other types of obligations namely at-decision obligations and pre-decision obligations. At-decision obligations should be enforced in parallel to the access (e.g., monitoring actions, notifications ...). Pre-decision obligations are to be enforced prior to the access (e.g., anonymization to reduce risk or second-factor authentication to increase the trust). Obligations in Chen's model is expected to be fulfilled by the data requester, whereas in our model obligations (all types) can be required from both data requester and/or the system (storing and controlling access and usage of the data) which allows more leverage to the adjustment strategy.

Using obligations as trust enhancement mechanisms has also been proposed by several other studies. In [7] we find a distinction between two categories of obligations: *provisions* or *pre-obligations* actions, which must be executed as a pre-condition for authorization decision; and *post-obligations* actions that must be fulfilled after the authorization decision is made. In [6], the trust value of a user is impacted by his previous history of fulfilling or not post-obligations, also considering their level of criticality. Access and Usage control (AC/UC) Obligations have been proposed to address privacy requirements and lower privacy violation likelihood in [2,15,35].

Following the original Chen et al. [10] model, these papers consider trust as part of the risk value. We can essentially map our model to this approach; in fact, by renaming the difference "$R - T$" as the "risk" in Eq. 1, and explicitly setting a threshold, we obtain mostly the same as described in Chen's model. However, as we discussed in [3], explicitly introducing the risk/trust comparison allows for: *(i)* trust enhancement and risk mitigation strategies are clearly separated, making easier to find an optimal set of strategies to increase access, keeping risk under control, *(ii)* trust thresholds are not dependent on the risk scenario, and, if we consider multiple risk factors, we can compare the overall risk with the trust. Our model addresses these issues, clearly separating trust aspects from risk.

8   Conclusions and Future Work

Security monitoring solutions, such as TDS, are becoming more sophisticated, and they consume large and diverse datasets (logs), containing personal and sensitive data. As a result, these systems should comprise access control models able to address complex privacy and utility requirements.

In this paper, we propose a risk-based privacy-aware access control approach (described in [3,4]) for a threat detection solution, where anonymization is dynamically applied to reduce the privacy risk and, possibly, combined with obligations to increase trust. Automatically applying specific anonymization strategies, in real-time, for each pattern, we showed how this model is able to provide a simple solution for investigating potentially harmful patterns, with a minimal privacy risk. In the cases where significantly reducing risk results in an excessive degradation of the quality of data, the model supports mechanisms of trust enhancement via enforceable obligations to access less-anonymized data. We also showed that the anonymization step does not impact the real-time performance of the systems for typical data set. Finally, we proposed an architecture and a declarative policy approach to realize our model.

We based our analysis on real TDS, using a small sample of typical patterns. A more extensive analysis is needed to be able to implement a robust solution. In particular, the parameter setting (risk thresholds) can be complex in presence of a large number of patterns. We also implemented a k-anonymity model for estimating privacy risk, and, although widely used k-anonymity has its own limitation, for example, in the presence of multiple overlapping data sets, it is well known that the k-anonymity condition cannot be fulfilled (lack of composability). Despite the relevance of this issue in general cases, this limitation is mitigated, in our particular case, by the fact that we deal with a very dynamic data and results of requests/patterns produce data views that rarely overlap. Other privacy models exist, such differential privacy, which could also be integrated into the risk framework (see [27]).

Our approach grantees a given level of privacy considering single requests and separate semi-trusted actors (in our case an operator, an administrator, or an advanced administrator). The risk model could be extended to cover, historical attacks (i.e., attacks that take advantage of the acquisition of a history of requests) and collusion attacks (i.e., attacks achieved through the participation of several actors).

Acknowledgments. The research leading to these results has received funding from the FP7 EU-funded project SECENTIS (FP7-PEOPLE-2012-ITN, grant no. 317387).

References

1. Ali, M., Bussard, L., Pinsdorf, U.: Obligation language for access control and privacy policies (2010)
2. Ardagna, C.A., Cremonini, M., De Capitani di Vimercati, S., Samarati, P.: A privacy-aware access control system. J. Comput. Secur. 16(4), 369–397 (2008)
3. Armando, A., Bezzi, M., Cerbo, F., Metoui, N.: Balancing trust and risk in access control. In: Debruyne, C., Panetto, H., Meersman, R., Dillon, T., Weichhart, G., An, Y., Ardagna, C.A. (eds.) OTM 2015. LNCS, vol. 9415, pp. 660–676. Springer, Cham (2015). https://doi.org/10.1007/978-3-319-26148-5_45
4. Armando, A., Bezzi, M., Metoui, N., Sabetta, A.: Risk-aware information disclosure. In: Garcia-Alfaro, J., Herrera-Joancomartí, J., Lupu, E., Posegga, J., Aldini, A., Martinelli, F., Suri, N. (eds.) DPM/QASA/SETOP 2014. LNCS, vol. 8872, pp. 266–276. Springer, Cham (2015). https://doi.org/10.1007/978-3-319-17016-9_17
5. Armando, A., Bezzi, M., Metoui, N., Sabetta, A.: Risk-based privacy-aware information disclosure. Int. J. Secur. Softw. Eng. 6(2), 70–89 (2015)
6. Baracaldo, N., Joshi, J.: Beyond accountability: using obligations to reduce risk exposure and deter insider attacks. In: Proceedings of the 18th ACM Symposium on Access Control Models and Technologies, SACMAT 2013, pp. 213–224. ACM, New York (2013)
7. Bettini, C., Jajodia, S., Wang, X.S., Wijesekera, D.: Provisions and obligations in policy management and security applications. In: Proceedings of the 28th International Conference on Very Large Data Bases, VLDB 2002, pp. 502–513. VLDB Endowment (2002)
8. Bezzi, M.: An information theoretic approach for privacy metrics. Trans. Data Priv. 3(3), 199–215 (2010)
9. Brickell, J., Shmatikov, V.: The cost of privacy: destruction of data-mining utility in anonymized data publishing. In: Proceedings of the 14th ACM SIGKDD International Conference on Knowledge Discovery and Data Mining, KDD 2008, pp. 70–78. ACM, New York (2008)
10. Chen, L., Crampton, J.: Risk-aware role-based access control. In: Meadows, C., Fernandez-Gago, C. (eds.) STM 2011. LNCS, vol. 7170, pp. 140–156. Springer, Heidelberg (2012). https://doi.org/10.1007/978-3-642-29963-6_11
11. Chen, L., Crampton, J., Kollingbaum, M.J., Norman, T.J.: Obligations in risk-aware access control. In: Cuppens-Boulahia, N., Fong, P., García-Alfaro, J., Marsh, S., Steghöfer, J. (eds.) PST, pp. 145–152. IEEE (2012)
12. Cheng, P.-C., Rohatgi, P., Keser, C., Karger, P.A., Wagner, G.M., Reninger, A.S.: Fuzzy multi-level security: an experiment on quantified risk-adaptive access control. In: IEEE Symposium on Security and Privacy, pp. 222–230. IEEE Computer Society (2007)
13. Ciriani, V., De Capitani di Vimercati, S., Foresti, S., Samarati, P.: Theory of privacy and anonymity. In: Atallah, M., Blanton, M. (eds.) Algorithms and Theory of Computation Handbook, 2nd edn. CRC Press, Boca Raton (2009)
14. Clifton, C., Tassa, T.: On syntactic anonymity and differential privacy. Trans. Data Priv. 6(2), 161–183 (2013)
15. Di Cerbo, F., Doliere, F., Gomez, L., Trabelsi, S.: PPL v2.0: uniform data access and usage control on cloud and mobile. In: Proceedings of the 1st International Workshop on TEchnical and LEgal Aspects of Data PRIvacy and SEcurity. IEEE (2015)

16. Dickens, L., Russo, A., Cheng, P.-C., Lobo, J.: Towards learning risk estimation functions for access control. In: Snowbird Learning Workshop (2010)
17. eXtensible Access Control Markup Language (XACML) Version 3.0, January 2013. http://docs.oasis-open.org/xacml/3.0/xacml-3.0-core-spec-os-en.pdf
18. Friedewald, M., Pohoryles, R..J.: Privacy and Security in the Digital Age: Privacy in the Age of Super-Technologies. Routledge, London (2016)
19. Fung, B.C.M., Wang, K., Chen, R., Yu, P.S.: Privacy-preserving data publishing: a survey of recent developments. ACM Comput. Surv. **42**(4), 4:1–4:153 (2010)
20. Ghinita, G., Karras, P., Kalnis, P., Mamoulis, N.: Fast data anonymization with low information loss. In: Proceedings of the 33rd International Conference on Very Large Data Bases, pp. 758–769. VLDB Endowment (2007)
21. Josang, A., Ismail, R., Boyd, C.: A survey of trust and reputation systems for online service provision. Decis. Support Syst. **43**(2), 618–644 (2007). Emerging Issues in Collaborative Commerce
22. Kaempfer, M. (2015). http://scn.sap.com/community/security/blog/2015/03/04/sap-enterprise-threat-detection-and-siem-is-this-not-the-same
23. Kohlmayer, F., Prasser, F., Eckert, C., Kuhn, K.A.: A flexible approach to distributed data anonymization. J. Biomed. Inform. **50**, 62–76 (2014). Special Issue on Informatics Methods in Medical Privacy
24. Kounine, A., Bezzi, M.: Assessing disclosure risk in anonymized datasets. In: Proceedings of the FloCon Workshop, January 2009
25. Lakkaraju, K., Slagell, A.: Evaluating the utility of anonymized network traces for intrusion detection. In: Proceedings of the 4th International Conference on Security and Privacy in Communication Netowrks, SecureComm 2008, pp. 17:1–17:8. ACM, New York (2008)
26. Li, X., Zhou, F., Yang, X.: A multi-dimensional trust evaluation model for large-scale P2P computing. J. Parallel Distrib. Comput. **71**(6), 837–847 (2011)
27. Metoui, N., Bezzi, M.: Differential privacy based access control. In: Debruyne, C., et al. (eds.) OTM 2016. LNCS, vol. 10033, pp. 962–974. Springer, Cham (2016). https://doi.org/10.1007/978-3-319-48472-3_61
28. Metoui, N., Bezzi, M., Armando, A.: Trust and risk-based access control for privacy preserving threat detection systems. In: Dang, T.K., Wagner, R., Küng, J., Thoai, N., Takizawa, M., Neuhold, E. (eds.) FDSE 2016. LNCS, vol. 10018, pp. 285–304. Springer, Cham (2016). https://doi.org/10.1007/978-3-319-48057-2_20
29. Mivule, K., Anderson, B.: A study of usability-aware network trace anonymization. In: Science and Information Conference (SAI), pp. 1293–1304. IEEE (2015)
30. Mont, M.C., Beato, F.: On parametric obligation policies: enabling privacy-aware information lifecycle management in enterprises. In: Eighth IEEE International Workshop on Policies for Distributed Systems and Networks, POLICY 2007, pp. 51–55. IEEE (2007)
31. Narayanan, A., Huey, J., Felten, E.W.: A precautionary approach to big data privacy. In: Gutwirth, S., Leenes, R., De Hert, P. (eds.) Data Protection on the Move, pp. 357–385. Springer, Dordrecht (2016). https://doi.org/10.1007/978-94-017-7376-8_13
32. Council of Europe: Handbook on European data protection law. Technical report (2014)
33. Committee on Strategies for Responsible Sharing of Clinical Trial Data: Sharing Clinical Trial Data: Maximizing Benefits, Minimizing Risk. National Academies Press, Washington, DC (2015)

34. Oprea, A., Li, Z., Yen, T.-F., Chin, S.H., Alrwais, S.: Detection of early-stage enterprise infection by mining large-scale log data. In: 2015 45th Annual IEEE/IFIP International Conference on Dependable Systems and Networks (DSN), pp. 45–56. IEEE (2015)
35. Pretschner, A., Hilty, M., Basin, D.: Distributed usage control. Commun. ACM **49**(9), 39–44 (2006)
36. Samarati, P.: Protecting respondents' identities in microdata release. IEEE Trans. Knowl. Data Eng. **13**(6), 1010–1027 (2001)
37. Sandhu, R., Park, J.: Usage control: a vision for next generation access control. In: Gorodetsky, V., Popyack, L., Skormin, V. (eds.) MMM-ACNS 2003. LNCS, vol. 2776, pp. 17–31. Springer, Heidelberg (2003). https://doi.org/10.1007/978-3-540-45215-7_2
38. Scholl, M.A., Stine, K.M., Hash, J., Bowen, P., Johnson, L.A., Smith, C.D., Steinberg, D.I.: SP 800–66 REV. 1. An introductory resource guide for implementing the health insurance portability and accountability act (HIPAA) security rule. Technical report (2008)
39. Shaikh, R.A., Adi, K., Logrippo, L.: Dynamic risk-based decision methods for access control systems. Comput. Secur. **31**(4), 447–464 (2012)
40. Templ, M., Meindl, B., Kowarik, A.: Introduction to statistical disclosure control (SDC). Project: Relative to the testing of SDC algorithms and provision of practical SDC, data analysis OG (2013)
41. Ulltveit-Moe, N., Oleshchuk, V.A.: Measuring privacy leakage for IDS rules. CoRR, abs/1308.5421 (2013)
42. Ulltveit-Moe, N., Oleshchuk, V.A., Køien, G.M.: Location-aware mobile intrusion detection with enhanced privacy in a 5G context. Wirel. Pers. Commun. **57**(3), 317–338 (2011)
43. Vaidya, J., Clifton, C.W., Zhu, Y.M.: Privacy Preserving Data Mining, vol. 19. Springer, Boston (2006). https://doi.org/10.1007/978-0-387-29489-6
44. XACML Obligation Profile for Healthcare Version 1.0, February 2013. http://docs.oasis-open.org/xacml/xspa-obl/v1.0/csd01/xspa-obl-v1.0-csd01.html
45. Zuech, R., Khoshgoftaar, T.M., Wald, R.: Intrusion detection and big heterogeneous data: a survey. J. Big Data **2**(1), 1–41 (2015)

Systematic Digital Signing in Estonian e-Government Processes

Influencing Factors, Technologies, Change Management

Ingrid Pappel[1(✉)], Ingmar Pappel[2], Jaak Tepandi[1], and Dirk Draheim[1]

[1] Large-Scale Systems Group, Technical University of Tallinn, Tallinn, Estonia
{ingrid.pappel,jaak.tepandi,dirk.draheim}@ttu.ee
[2] Interinx Ltd., Tallinn, Estonia
ingmar@interinx.com

Abstract. In Estonia, digital signing started with the Digital Signatures Act already as early as in 2000. The aim to make digital signing and its use with various types of documents more convenient and efficient has had a high priority in the state's e-Governance initiative. In this article we provide a study of the systematic introduction and use of digital signatures with documents related to decision-making processes and analyze the factors which influence this. We look at local governments as a major use case and provide an overview of the digital signing statistics for local government document exchange. The article highlights the differences related to the size and administrative capacity of the local governments as well as their readiness to transition into the information society.

Keywords: Digital signing · Digital document exchange
Digital administration

1 Introduction

We live in an increasingly digitalized world. In addition to the different technological solutions in everyday life, document management and the related decision-making processes have also become digital. The Digital Agenda for Estonia 2020 aims for a "simpler state" [1], whereby in order to make the public sector more effective, it is important to achieve a 95% paperless official communication rate by 2020. This requires local government services to be as electronic as possible and that as an end result of the provided services, instead of printing out a paper to prove the fact of service provision, it is stored in digital form. In order to achieve this, various procedural systems are in use in Estonia, including document management systems (DMS) – which comprises of and manages documents as well as facilitates constant access to them. DMS has brought transparency to administration and allowed for including citizens in the decision-making processes of the organization. This, in turn, has made the implementation of digital signatures more efficient in Estonia.

In this paper the correlation between the use of digital signatures and specific document types is discussed based on usage of the DMS Amphora. Additionally, a

© Springer-Verlag GmbH Germany 2017
A. Hameurlain et al. (Eds.): TLDKS XXXVI, LNCS 10720, pp. 31–51, 2017.
https://doi.org/10.1007/978-3-662-56266-6_2

survey has been conducted that provides an overview of the factors influencing digital signing in local governments. Various research methods were used to carry out this survey, such as data obtained from databases on the basis of specified criteria, the observation of world practices, questionnaires and interviews. Generalizations have been made based on more than 50% of the Estonian local governments.

In order to make digital document exchange more efficient, several solutions have been developed in Estonia [2], e.g. the document exchange center (DEC) and e-services at the citizen portal eesti.ee environment, which enable the digital processing and management of a document life cycle from its birth to death. Over the years, the volume of paper documents exchanged between authorities has decreased significantly [3], which in turn has a positive effect on the budget of the institution. DMS Amphora is used in 127 Estonian local governments and this article presents the data from 117 Estonian local governments because their data was available in the database in the proper form. The data has been taken about the first quarter of 2016. The software solution enables to observe the reply deadline for the letters, and to digitally sign all documents and letters. The data used in this work have been obtained from the DMS database according to the following:

- How many incoming documents has the given local government registered in the Amphora document management system;
- How many outgoing documents has the given local government registered;
- How many of the outgoing documents has the given local government signed digitally;
- Total numbers of letters and documents;
- Capability index of the local government units [4];
- Number of residents in the given local government;
- How many documents per residents are there in the document management systems in the first quarter;
- Name of the local government;
- County in which the local government is located.

In Sect. 2 we explain the background, i.e., the current state of digital signing in Estonia and its motivation. Also we report on first insight concerning problems with digital signing and digital archiving. In Sect. 3 we give an overview of the technological infrastructure in Estonia relevant to digital signing. In Sect. 4 we provide the results of a survey concerning digital signing. In Sects. 5 and 6 we derive factors and draw recommendations for digital signing from the survey results. In Sect. 5 we delve further in issues of change management for the introduction of digital signatures in organizations. In Sect. 7 we discuss possible strands of future work. Related work is discussed in Sect. 8. We finish the paper with a conclusion in Sect. 9.

2   Digital Signing in Estonia

As aforementioned before, the digital signatures in Estonia is governed by the Digital Signatures Act (DAS), which was adopted on 7 March 2000 [5]. In the eyes of the law, a digital signature is equal to a handwritten signature. All Estonian authorities are required to accept digitally signed documents. Estonian public authorities are required to accept digitally signed documents. Two certificates are issued along with an ID-card. One certificate is for identification and the other for digital signatures. It is important to ensure that these certificates have not expired when using digital signatures. In addition to signing by using the ID-card, Mobile-ID signatures are becoming increasingly popular. In 2015, the number of Mobile-ID users increased by 40%, exceeding the 75,000 user line this January. These users carried out over 25 million Mobile-ID transactions in the last year. In 2014, Mobile-ID was used for an average of 1.8 million transactions per month, where 2015 the monthly average was already 2.7 million [6]. Three types of formats are used in Estonia – BDOC, DDOC, and CDOC. The oldest one of these, the original is the DDOC. BDOC is a newer format meant for replacing the DDOC format, and it is certainly more consistent with international standards. CDOC is a file which in its encrypted form contains a data file (XML document or other binary file, e.g. MS Word, Excel, PDF, RTF, etc.), the certificate of the recipient, an encrypted key for data file decryption, and other optional metadata [6].

2.1 Reasons for Using Digital Signatures

A digital signature is the counterpart of an ordinary signature used to sign information in digital form. Digital signatures help identify the link between the document and the person who signed it. A digital signature along with a time stamp forms a combined dataset with the document, the components of which cannot be individually altered at a later time. Digital signatures replace ordinary signatures which helps to ensure the authenticity and security of electronic documents. Besides apply paperless administration to enable digital document exchange [7]. Ensuring security with a digital signature means that the document author is known and the document has not been altered by third parties between being sent and received [8]. The digital signature standard (DSS) was created by the US National Security Agency. DSS is based on the digital signature algorithm (DSA). DSS can only be used for digital signatures but the DSA can also be employed for encryption [9]. The simplicity of digital signing can be considered its biggest advantage. It is quick and convenient and lacks many of the risks that signing on paper entails. It is certain that a physical person is responsible for the signature. The signed document has not been subsequently edited by third parties, this option is eliminated by mathematical links. It is always possible the check the signing date because the time stamp is a part of digital signing.

- An endless number of legally equal copies can be made of a digitally signed document.
- Digital documents do not take up physical space.
- Digital documents do not require paper, a printer or other superfluous resources.
- Digital documents do not need to be delivered and communication is possible through electronic channels.
- With the use of DMS, digital documents can be found more quickly and archived on the basis of very different criteria.

When signing digitally, one must consider that the generated file can be singly read using convenient methods by all interested parties and that it can be opened without issues in the future as well. If a file has been signed in one format, then it cannot be converted into another format without losing the signature. It is important to use to correct file formats for signing so that the file meets all the requirements. There are several possible purposes for using digital signatures, e.g. no need to specifically meet in person for a signature or to send documents with ordinary mail, thus significantly saving time. Digital signing allows for automating activities and to reduce spending time on regularly signing a large number of documents physically. If necessary, the document should be encrypted so strangers cannot read it.

2.2   Problems Related to Digital Signing

From a local government perspective, several issues have been highlighted that are related both to the organizational as well as technical aspects. Also, this is widely discussed elsewhere as well [10]. From a technical point of view, the digital signature format can be limited, as it is possible that different environments can show the document in different ways. The most important and serious risk with using digital signatures is that the signature rights can be stolen with a private key – the owner of the certificate must carefully monitor that the private key does not leave the possession of the signature owner. Nowadays, different methods have been devised to tackle this and the risk is diminishing.

The problems that may arise when using digital documents tend to differ between small- and large-scale uses. In both cases, one must bear in mind that not all clients and partners may have an ID-card or Mobile-ID and that parallel paper document use must be retained. The latter can only be avoided when an authority issues unilaterally signed documents. This could create duplication. For small-scale use, e.g. internal use of an organization and signing contracts with larger partners and clients, different issues occur and the use of a computer and ID-card and passwords is an extra effort, takes more time and is not suitable in outdoor conditions. In addition, a problem with digital documents may arise regarding the accompanying time stamp – the physical time of signing is visible to everyone who looks at the document. In local governments, this is linked to certain decisions and the granting of rights, where an important administrative act is formalized after the fact.

2.3   Problems Related to Archiving

Many local governments have brought out archiving as an issue for digital signing. Archiving digitally signed documents requires some extra effort [10–12]. With archiving, one must take into account that in addition to digital documents, paper documents also need to be managed. Thus, hybrid files are created. Inevitably, it is more difficult to use two separate management systems rather than only have one; it is reasonable to manage both digital and paper documents in the same information system. A solution is that the location, existence, and main information (what type of document, what parties, when, etc.) about the paper documents is registered in the same information system and in the same way as for digital documents, in the simplest case by using a small ordinary document file containing the main information. If an organization already employs a paper document registration system, adding digital document management to the same system is likely to be the most effective – provided that this is technically possible.

Regarding potential software solutions, it is important to consider whether an existing software already in use could be suitable for archiving digital documents, or if the standard activities used in the organization already could not be employed for archiving the files. For instance, digital documents could simply be stored in a file system, grouping them chronologically into year- or month-based catalogues and coding the critical information (client name or code) into the file name. This can be used if there is a relatively small amount of digital documents. In addition, an existing specialized archiving software could be used and generally, a DMS already contains an archiving function. If it does not, a suitable archiving software could be created for the organization.

2.4 Digital Signatures and Digital Document Authenticity

Is a digital signature always a sufficient guarantee of the digital document authenticity for digital archiving? From the perspective of the Estonian national archive, it can be not sufficient [13–15]. A digital signature does protect the signed information (the content of the document) from unwanted changes but it is not enough to completely understand the document. A part of the information no less important than the content is hidden in the links between the documents – these allow us to understand the activities of the organization, during which the document was created. A digital signature does not release an organization from good and controlled management of the document, which is one of the guarantees of document authenticity. In the case of signed, but especially for digital documents with a permanent retention period, the organization must implement and ensure specific policies and procedures that enable verifying the creation, sending, forwarding, retention, and separation of documents [14].

In the future, it is possible to use archival time-stamping for ensuring the long-term preservation of documents in the BDOC format. This mechanism is based on the principle "fortify that, which could be weak" [13]. Consecutive time stamps protect the entire contents from weak hash algorithms and from breaching cryptographic material and algorithms. Certain costs are associated with this, as there is a need to enter into a contracts with an organization that offers certification and time stamping services (presently, in Estonia, this organization is Certification Centre). Monthly bills also need to be paid for the validity confirmation service, however, the costs are not that big.

3   Technological Infrastructure

In this section we give an overview of the Estonian technological infrastructure that enables digital signing of documents as represented in this article, compare also with [16]. Digital signing of documents is two-tiered. Each document is signed organizationally by Estonian's streamlining data ex-change backbone, then, each document is signed by the person who is the accountable stakeholder in the respective organizational process. It is the latter, the individual signature, that we treat as digital signing, the first can then be coined e-Stamping.

In Estonia, e-Government is enabled and streamlined by a systematic distributed architecture and infrastructure, often coined X-Road, compare also with [17]. X-Road is called a data exchange layer, but is way more than just a data exchange layer protocol. It is the entirety of organizational and technological assets that enable a secure, tamper-proof and repudiation-proof data exchange over the public internet. Between parties involved in e-Governance processes, compare with Fig. 1. Basically, X-Roads consists of a data exchange layer protocol based on SOAP, the specification and implementation of an organizational security server, a PKI (public key infrastructure) that shows, in particular, in trust services for certificate validation and time stamping plus procedures for registration of X-Road members, organizational security servers and data services plus regulations for the establishment of organizational data bases in the X-Roads environment. The idea is that all e-Governance is streamlined by X-Road. Organizations that want to take part into Estonian's e-Governance need to become members of X-Road and must adhere to its standards and regulations. X-Roads follows a lightweight, distributed approach that aims at keeping centralization at a minimum. The key principle is that organization keep ownership and responsibility of exchanged data. Therefore, security servers are run by the single X-Road members.

Messages are sent directly from one organization to another via the organization's security servers. This means, that X-Road is not a value-added network, not an ESB (Enterprise Service Bus) no message-oriented middleware (MOM) or the like, compare also with [18–21]. The security servers take care for encrypting/decrypting, e-stamping, validating and time stamping outgoing and incoming messages. They exploit regulated trust services for that purpose that are provided by third party certification authorities. These trust services are a certificate validation service based on OCSP (Online Certificate Status Protocol) and a time-stamping service. The X-Road specified trust services adhere to the EU eIDAS regulation on electronic identification [22]. Each organization keeps full control over the data in its databases and connected information systems, in particular, it is the single organization that grants access rights for their data

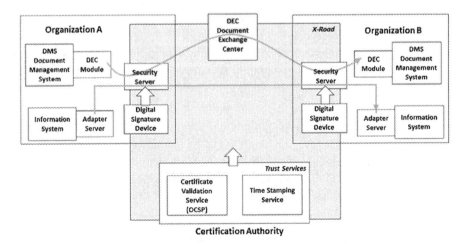

Fig. 1. Data exchange layer X-Road with trust services and document exchange center

to other organizations. Each organization maintains and controls these access rights in its own security serve. Each message exchanged via X-Roads is digitally signed by the security server of the sending organization.

For the exchange of documents an additional service, the document exchange center (DEC) has been established. The DEC slightly deviates from the X-Roads key principle of minimal centralization. If provides a store-and-forward mechanism therefore a kind of enterprise service bus component. Documents are sent to DEC where they are temporarily stored and preserved. Organizations that are entitled to, can than pick documents for the DEC. Organizations that want to participate in document exchange via DEC must be registered members of XROAD, in addition, they must become also registered users of the DEC.

4 Digital Signature Statistics Based on DMS Databases

The study presented in this article only reflects the digital signing of documents exchanged using the DMS, but many documents are processed outside of the document management system using other components [17]. For instance, if one were to change one's place of residence and make an application about this to the local government, this application is registered as an entry in the Population Register and may well not be reflected in the document management system. The same applies to construction permits, authorizations for use, and applications for design criteria, which are all registered in the Construction Register. The data that are registered in the social services and benefits data register (STAR) are also excluded from the document management system. In Estonia, information exchange with other systems is mainly carried out over the X-road for relational systems [16]. However, this is not always the case, and therefore it is necessary to also observe the situations where information with external systems is exchanged outside the X-road, in order to have adequate statistics about the public sector document exchange. Although X-road is the preferred communication channel, there are still information systems that communicate directly, i.e. exchange documents by other interfaces. Below, data is shown in various groups (local government totals, more successful local governments, less successful local governments, etc.), bringing out volume of digitally signed documents (Fig. 2, Tables 1, 2, 3, 4 and 5).

5 Factors Influencing Digital Signing

During the application of paperless management, there are several factors that determine its success. Same applies to digital signing as it is one important part of DMS-s. Although this paper discusses results based on survey in 2016, there have been other experiments before. The study was conducted in Rapla County during 2009–2011 [23, 24] also showed that local governments need a solution that would unify their services. After taking e-forms into use, an increase by leaps and bounds in digital signing was also evident (see Figs. 3 and 4) as the procedural steps of the respective applications were performed digitally and the answers to citizens were also transmitted digitally [24].

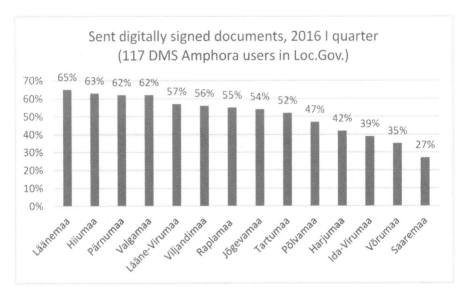

Fig. 2. Summarization according to counties

Table 1. Consolidated data

Consolidated data	
Total number of sent documents	24801
Total number of digitally signed sent documents	12245
Percentage of digital signing for sent documents	49%
Number of local governments in the sample	117
Average number of residents in local governments in the sample	3298

Table 2. Local governments that use digital signing the most

Local government	County	Sent signed	Number of residents	Capability index ranking
Tori parish	Pärnu	86%	2327	127
Elva town	Tartu	85%	5768	39
Värska parish	Põlva	81%	1374	113
Tahkuranna parish	Pärnu	81%	2389	114
Audru parish	Pärnu	81%	5858	52
Karksi parish	Viljandi	80%	3400	88
Vigala parish	Raplamaa	79%	1267	66
Paikuse parish	Pärnumaa	75%	3899	74
Kehtna parish	Raplamaa	75%	4459	49
Vinni parish	Lääne-Viru	75%	4757	21

Table 3. Local governments that use digital signing the least

Local government	County	Sent signed	Number of residents	Capability index ranking
Pihtla parish	Saare county	2%	1411	109
Ahja parish	Põlva county	0%	1011	191
Kihelkonna parish	Saare county	0%	773	86
Laimjala parish	Saare county	0%	711	183
Meeksi parish	Tartu county	0%	594	194
Mustjala parish	Saare county	0%	691	207
Sõmerpalu parish	Võru county	0%	1799	118
Torgu parish	Saare county	0%	350	208
Torma parish	Jõgeva county	0%	1991	137
Varstu parish	Võru county	0%	1075	180

Table 4. Most digitally signed letters per resident in local governments with up to 10,000 residents

Local government	County	Sent per resident	Number of residents	Capability index ranking
Lüganuse parish	Ida-Viru	0.235	3014	23
Vihula parish	Lääne-Viru	0.117	1955	36
Piirissaare parish	Tartu county	0.098	102	210
Vormsi parish	Lääne county	0.096	415	75
Misso parish	Võru county	0.096	645	126
Meremäe parish	Võru county	0.092	1093	181
Mõniste parish	Võru county	0.084	873	166
Kernu parish	Harju county	0.084	2040	27
Värska parish	Põlva county	0.080	1374	113
Are parish	Pärnu county	0.069	1297	122

The implementation of e-services in the governing arrangement of local governments has a positive impact, because it facilitates solving the issues of the citizens more operatively and permits the better monitoring of the whole course of proceedings.

Table 5. Number of digitally signed letters per resident in local governments with more than 10,000 residents

Local government	County	Documents per resident	Number of residents	Capability index ranking
Viimsi parish	Harju county	0.010	18430	4
Viljandi town	Viljandi county	0.028	18111	32
Rae parish	Harju county	0.035	15966	1

(continued)

Table 5. (*continued*)

Local government	County	Documents per resident	Number of residents	Capability index ranking
Rakvere town	Lääne-Viru county	0.021	15942	40
Maardu town	Harju county	0.005	15676	29
Saue parish	Harju county	0.032	10451	7
Haapsalu town	Lääne county	0.037	10425	41

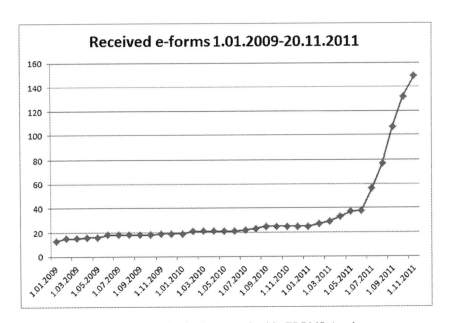

Fig. 3. The growth of e-forms received in EDRMS Amphora

5.1   Outcomes of the Survey

In this section we delve into the factors influencing digital signing by seeking for generalizations based on survey. This analysis is based on the survey conducted in spring 2016, which examined the various factors that influence the implementation of digital signing in local governments. The answers obtained from the survey illustrate the main factors which obstruct or advance digital signing in DMS. The answers reflects different criteria and measurements sets concerning the digital signing. For instance, answers to the questions "Do you sign government legislation digitally" the "yes" was answered 39,3%. Question "Do you sign outgoing documents digitally" got 58,2% "yes" and "partially" 40%. "Do you think preserving digital signatures is safe?" gave 46,4% "maybe" and 49,1 "yes". Question "Do you think digital signatures can be used as evidence (e.g. in court)" got 79,8% "yes" answers. To the question "Is forwarding

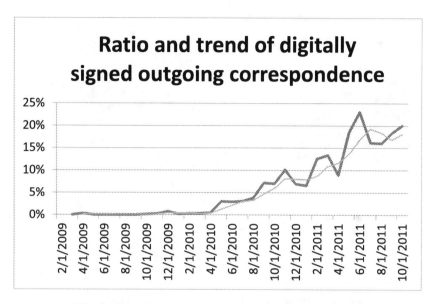

Fig. 4. The increase in digitally transmitted correspondence

digitally signed documents to citizens an issue" gave 57,4% of "Yes answers". On the following figures are shown different criteria which were investigated such us variety of age and different factors influencing the digital signing (Figs. 5, 6, 7 and 8).

Also, in the inquiry, there was an open text question "What should be done to introduce the digital signing in depth". The most used suggestions were brought out as follows:

- In order to raise elder people capability, the access to a computer, internet should be guaranteed more widely
- Digital signing should be introduced (forced) by rural municipality mayor within organisation (local government)
- Raise awareness regarding the digital archiving – explain long-term preservation methods
- It is necessary introduce and market digital signing for both - officials and citizens
- Develop more Public Internet Access points (for instance use county's library), which gives the opportunity to consume public e-services (different application)

Allover, from the survey, we learned that the following are the delaminating factors for digital signing:

- *Digital Divide*
    - elder people vs. younger people
    - lack of ubiquities internet access
- *Lack of sponsorship*. Lack of sponsorship by leaders in administrations.
- *Lack of awareness concerning digital archiving.*
- *Lack of iniquitousness towards population.* Barrier in the usage of digital signing between officials and citizens.

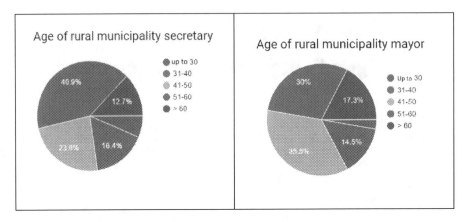

Fig. 5. Age difference

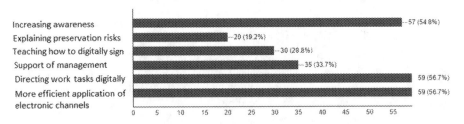

Fig. 6. How to raise digital signing?

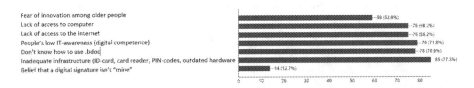

Fig. 7. Please mark the factors which could prevent forwarding digitally signed documents to citizens

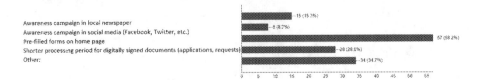

Fig. 8. How do you motivate your citizens to use digital signatures?

5.2 Organisational Development and Change

From the previous chapter it occurred that there are many factors influencing digital signing. The Digital Divide was stressed by drawing a gap between older and younger people habits, also lack of ubiquities internet access was pointed out. However, an important factor was brought out by not supportive management, and also awareness regarding the digital preservation. Also, implementation of digital signing faces controversies between the requirements arising from static legislation on the one hand, and the use of progressive ICT tools on the other hand. All these factors can be more or less related to the organizational development and change management in general. Organizational Development (OD) aims to expand the knowledge and effectiveness of people to accomplish more successful organizational change and performance [25].

OD is a process of continuous diagnosis, action planning, implementation and evaluation, with the goal of transferring knowledge and skills to organizations to improve their capacity for solving problems and managing future change. According to French and Bell (2011) organization development (OD) can be defined as "organization improvement through action research" [26]. During the 2003–2016 the developments of DMS Amphora as whole were mirroring the same logics where outcomes were constantly evaluated, improved and by that initiated a new cycle of investigations (see Fig. 9).

In addition to development activities of DMS Amphora the implementation process itself has been highly considered as a tool for managing organisational changes. Organisations have their own culture and specific ways do things. Especially in public sector government offices, there everything is strictly established based on legal environment and state functions. Alongside with the legal obligations many unwritten rules occurs that nobody is consciously aware of, still these dictate many decisions. All cultural habits have to change if digital transformation is going to take hold over the

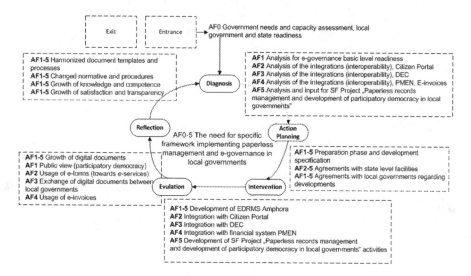

Fig. 9. Research and development activities of DMS Amphora

long term. But overcoming obstacles and new sometime unwritten rules is not easy. However, a fear can also hinder a progress as for someone digital signing is the unknown, and for many term digital is a big unknown. Many people afraid it will make their role redundant and people fear to learn new skills while analyzing their capacity to do that. From the survey it occurred that the factors which could prevent forwarding digitally signed documents to citizens were mostly related to people's low IT literacy (71,8%) and that people don't know how to use the format.bdoc (70,9%). Also the compatibility of the software-hardware was highly mentioned (85%) but it could be related to the IT literacy in general. It is impossible to overcome people's fears if there is a lack of communication. If people are struggling with the unknown, there is a need to make the unknown as known. Therefore a great deal of regular and consistent communication will helps to overcome this gap. During the years there have been conducted several implementation stages in local governments since 2009 [2, 23, 24].

Conducting changes is only possible by management exemplifying and communicating a new reality over the long term and doing it constantly. Set a new direction and step back does not work, management needs to remain engaged with the process. Another fact which may slow down change is related to novelty and obscurity in order to implement new way of thinking. According to Calista et al. [27] early adopters of the digital government often found it difficult to maintain their performance, while some late adopters have experienced dramatic performance improvements. Total implementation of the digital signing takes much effort and in many cases the business processes should be re-engineered. Development of the harmonised implementation methodology for local governments to use DMS Amphora has been an important goal for years. In the beginning of 2011, the objective was to partially develop the methodology for increasing and measuring the digital performance of local governments on the basis of nine local governments of Rapla County – the project was called "e-Raplamaa" and some results were presented at an e-Governance conference in Tallinn [23]. An important objective of this project was to create the methodology and criteria for measuring the changes in proportion and effectiveness of digital administration in local governments. One of the aims was to give an answer to whether and how much the training and application of DMS increases the proportion of digital administration in local governments. The statement was that the effectiveness increases at least 20% in a three-month period after the application and, thereafter, there will be no increase, i.e. the growth stops. In order to prove that two application/training days were organised in each local government. In conclusion the collected results did not give sufficiently adequate feedback to draw direct connection between trainings and efficiency growth in order to make correlation. Still, it helped to develop the Digital Performance Index [23]. Based on these results was generalized that local governments, who have had more training days and are using DMS functionality widely are also more effective users of the digital signing. In addition to every-day documents and records management, the head of the organizations have to understand the responsibilities which are related to digital preservation. As it was seen from the survey there is still lack of awareness concerning digital archiving.

5.3 Raising Awareness on Digital Preservation

The conducted survey demonstrates the necessity to raise awareness regarding the digital archiving by explaining long-term preservation methods. It is clear that the advantage of the electronic records is that they are reusable. Also, it is possible very quickly adapt a record or compile a new record on the basis of an existing one. It can be digital advantage or at the same time vulnerability because adaptations or changes are not always observable and retained. Still, the efficiency and time saving on the digital workflows is worth to implement. Thus, local governments can raise ICT capacity by raising the awareness of their officials. Besides, the growth of citizen satisfaction is tied to the growth of the digital performance of local governments [24]. The implementation of digital document work proceedings is facilitated by the rules and instructions described on the state level wherein several problems still require solutions in order to reach a wider assessment of the synergies and cooperation between local governments and the state. These rules and standards should be explained more thoroughly local governments officials. Also, more trainings are needed. These training can be conducted online based as well. One of the main focuses during the development of DMS Amphora has been a common implementation methodology for effortless organisational change [28]. It provides an efficient learning environment for the users for understanding the functions of local governments and actions with digital environments like DMS and integrated systems. One part of that is digital archive module. The developed implementation methodology has helped to increases the awareness of the users by providing a common ground for understanding the benefits. In addition to described methodology there has been developed common e-learning environment for local governments [29]. It contains information (instructions, training videos, etc.) about DMS interface and functions, putting it into use and managing various functions. Besides, e-learning environment enables users to measure the level of their knowledge, skills and user experience. And, users can exchange their experiences in e-learning environments. Therefore, the proposed approach focuses largely on the pre-generated environment and process-based tutorials in an e-learning environment in order to train users on DMS Amphora. With some minor adjustments, the same methodology could be implemented on other DMSs as well [29]. Which in turn helps to raise awareness concerning the digital preservation as well.

6 Recommendations for Implementing Digital Signatures

In order to implement the digital signing efficiently, it is necessary to consider the most suitable scenarios regarding saving/archiving documents. In most cases, there is no need to develop or implement some kind of special software, and using freely available standard software will suffice. Still, the use of DMS allows for digitalising the processes inside the institution and, thus, is one of the most popular inter-governmental services in e-Government projects [30, 31]. Implementing digital signatures in the DMS requires certain changes which can be divided into organizational and technological. On the information technology level, the work of users should be made convenient and where possible, automatic storage of deliveries and work task flows should be introduced, as these support digital signing. On the organizational level, potential activities are mainly linked with training and increasing user awareness, both regarding simplifying work flows and digital archiving. If automatic work flow simplification is not possible, awareness campaigns are required and the users need to be taught how to a document is sent to be digitally signed when registered in the DMS or forwarded from the DMS. Digital signing is closely related to the implementation of digital records management. If the work processes are digital, digital signing is one logical step in the whole process. When discussing the digitalization of processes, it is important to note the complexity of business processes which in many cases are related to the size of the organization and the complexity of the offered services. For organizations with a rather large number of users, the complexity and large amount of business processes will be the deciding factors, nevertheless the people signing digitally tend to be the leaders of the organizations.

It is important to consider mapping, selecting, and analyzing the business processes suitable for the paperless alternative. The work load ranges from a few days to half a year, depending on the organization and the complexity of the task. Vitally, the preparation and carrying out of archiving digital documents must be planned. If the organization already has the required software or experience of using ordinary software, developing the principles for digital archiving is going to be easier.

Transitioning to digital signing in Estonia is also supported by a European Commission directive eIDAS [22]. The standards listed in this directive also include the bdoc-format digital signature used in Estonia. European public authorities are required to recognize digital signatures that meet this standard, thus providing an Estonian citizen with the right to bring an action against someone in a court in Barcelona that is signed digitally. On the other hand, Estonian public authorities have to learn to receive other types of digital signatures received from Europe. Estonian digital signatures must start accepting digitally signed documents with an equal or "stronger" signature from other European Union countries. Estonian citizens in turn get the opportunity to turn to other European public authorities with their digitally signed documents.

7   Future Work

The implementation of digital signatures is a key enabler for e-Government initiatives, similarly it is at the core of the paperless office. A stable infrastructure for digital signatures consists at least of a convenient to use public key infrastructure. Convenient to use means that the public key infrastructure comes with well-defined, transparent and available routines for registration, certification as well as services for validation. If it is expanded to quasi-standardized automatic interfaces to information systems, even better. If it is further expanded to quasi-standardized features in end-user tools, once more, even better. Once a stable infrastructure for digital signing is established, it enables the transformation of e-governance processes into purely digital processes. This is so for the realm of e-Government as well as e-Commerce. Here comes the point: whenever an organizational process reaches a certain criticality, a certain level of compliance relevance, we can almost be sure that some signing of documents is involved.

Implementing digital signatures immediately increases efficiency. However, it does not guarantee at all an improvement of the effectiveness. The implementation of digital signatures can be done, and this is actually the most usual case, in a non-disruptive manner with respect to the existing processes. As a result, processes are completely digitalized, however, they are themselves not changed essentially. Here is where a next wave of enactment and enablement is possible, both inside organizations as well as cross-organizational, compare also with [32]. Here, digital signing is really just the basis, albeit an essential one. The real efforts are in the assessment and re-design of existing processes, a huge refactoring and change management endeavor, when it comes to the cross-organizational cases, which are actually the most interesting ones, i.e., the ones with the highest potential to increase effectiveness. We have started basic, use-case driven, research in this direction. The point is that we need to start from scratch, even in some basic cases, and need to conduct system analysis. Currently we investigate how to exploit best practices, techniques and tools from the realm of enterprise architecture (TOGAF, DODAF, Zachman framework) [33–35] in the analysis and refactoring of cross-organizational administrative processes. There is also potential for innovative supporting tools, like cross-organizational business activity monitoring. As a concrete next step, we investigate how to integrate a business rule engine into the document ex-change center DEC.

Also with respect to supporting business process technologies [36], there are still many opportunities. To see this, we start with identifying two different kinds of qualities of digitally signed documents. The first kind is, what we would like to call asset-related, the second is what we would like to call process related. Asset-related documents serve as proof of ownership or right. They are independent of particular organizational processes, albeit they play crucial roles in organizational processes over and over again. Process-oriented signatures stem from the organizational processes themselves. Organizational processes emerge and are shaped over the years; many of are built around some RACI principle (responsibility, accountability, consulted, informed). Then digitally signed documents have the purpose to allow for next activities. Traditionally, they serve as a message, a trigger so to speak, but also have the purpose of documentation and proof, two facets that are important with respect to compliance issues. At least, with respect to the kind of process-related documents we should think about their transformation into digitally signed workflow steps and work-flow triggers. Which leads us to a vision of signature-integrated workflow management system. A similar vision is currently developed by the smart contract community, starting from a particular cross-organizational perspective, compare with [37, 38].

8 Related Works

Digital signing Problems related to digital signing are widely discussed from the perspective of the integrity and authenticity [10], and digitally signed documents requires extra effort for digital archiving [11, 12, 39]. In order to guarantee the organization in Estonia must implement and ensure specific policies and procedures [14], besides the initiative comes from the EU level as well. However, investigating other countries experiences several circumstances indicates the rise of the digital signing. Levy [40] recognizes that *"to benefit from its massive advantages, digital signatures still have challenges to overcome"*.

According to [40] the financial services industry has been the pioneer in the adoption and development of digital signature solutions, and he expects other industries, such as telecommunications, commerce, utilities, notaries and healthcare, to follow suit. Estonian case shows that besides the financial service industries the public sector has been adopted digital signatures quite well as well. However, based on the report [40] the findings are claiming that *"challenges include the integration and alignment of the technology with existing processes, together with a transparent analysis of the related regulatory situation and its legal consequences when implementing digital signatures"*. On this basis, it should be admitted that same matter must be considered in Estonian case. Although, the digital signatures are more efficient way to work, still the different obstacles should be resolved first. Besides the legal framework, the problems related to digital signing are tight to technology issues and people's resistance. This is discussed in the study conducted in USA where survey [41] shows that *"digital signatures have emerged as one of the technology priorities for local and state governments for the purpose of gaining both operational efficiencies and legal assurances"*. Like to this paper, the aforementioned survey was conducted among the local and state authorities and shows many similarities in findings to this work here as well. Still, the main advantages of this presented work are presenting besides the qualitative research results based on statistics from the DMS databases. This in turn gives real-live numbers of the actual signing of the local governments and qualitative research helps to understand the difference of the curve within local governments. To conclude, the international studies are indicating that digital signing is an important future trend and its development should be considered, while making local governments work routines more efficient along with the cost savings on paper products.

In [42] we report on the implementation of e-Invoicing in Estonia, again based on a document management systems approach. The described approach follows stepwise enterprise application via workflow modules and interfacing with enterprise resource planning (ERP) systems.

In [43], the authors demonstrate how to integrate digital signature workflow management into enterprise content management systems, based on secure digital tokens via smart cards. Documents can be signed each smart card having digital signatures capabilities with the citizen card as a particularly important case. In [44] the authors report on intrinsic barriers of the implementation of digital signatures. The Russian e-Government initiative is used as a case study for this purpose.

9   Conclusion

Digital signing has already claimed a significant place in today's society but signs are showing that the importance of digital signing is bound to increase even more in the near future. Firstly, the simplicity and security of the signature make it a preferred choice ahead of signing on paper. Secondly, digital document exchange also translates into savings in the budget. It can also help increase the security of the documents: a digital signature is tamper-proof and creates the option of creating an unlimited number of authentic verifiable copies of the document. This in turn enables to reduce the work load and increase the efficiency of local governments. Although the survey revealed that many of the smaller local governments do not have such administrative capabilities, the proportion of digital signatures is still notable. The main findings of the survey can be summarized as limiting factors concerning digital signing, which are digital divide, lack of sponsorship, lack of awareness concerning digital archiving and lack of iniquitousness towards population. For more efficient implementation, in addition to technological adaptations, the awareness of officials about issues related to digital archiving as well as software capabilities and interoperability for reading documents should be increased. Thus, in order to improve the implementation of digital signing in an organisational setup, there is a need to increase the IT-literacy of local government officials. The latter leads to the better management of the organisational changes and helps beside the digital signing more efficient digital workflow.

References

1. Ministry of Economic Affairs and Communication: Estonian Digital Agenda 2020, Tallinn (2013)
2. Pappel, I., Ingmar, P.: Implementation of service-based e-government and establishment of state IT components interoperability at local authorities. In: The 3rd IEEE International Conference on Advanced Computer Control (ICACC 2011), Harbin, China (2011)
3. Ministry of Economic Affairs and Communication: Outcome of the 2014 Estonian document exchange classification Project DECS, Tallinn (2015)
4. Ministry of Interior of Estonia: Classification of Estonian administrative units and settlements 2015v1, Tallinn (2015)
5. Estonian Digital Signature Act, Tallinn (2000)
6. Estonian Certification Center: Usage of ID in Estonia, Tallinn (2016)
7. Pappel, I., Pappel, I., Saarmann, M.: Digital records keeping to information governance in Estonian local governments. In: Shoniregun, C.A., Akmayeva, G.A. (eds.) i-Society 2012 Proceedings: i-Society 2012, 25–28 June 2012, London (2012)
8. Merkle, R.C.: A certified digital signature. In: Brassard, G. (ed.) CRYPTO 1989. LNCS, vol. 435, pp. 218–238. Springer, New York (1990). https://doi.org/10.1007/0-387-34805-0_21
9. Naccache, D., M'Raïhi, D., Vaudenay, S., Raphaeli, D.: Can D.S.A. be improved? — Complexity trade-offs with the digital signature standard —. In: De Santis, A. (ed.) EUROCRYPT 1994. LNCS, vol. 950, pp. 77–85. Springer, Heidelberg (1995). https://doi.org/10.1007/BFb0053426
10. Vigila, M., Buchmanna, J., Cabarcasb, D., Weinerta, C., Wiesmaier, A.: Integrity, authenticity, non-repudiation, and proof of existence for long-term archiving: a survey. Comput. Secur. **50**, 16–32 (2015). Elsevier
11. Lekkasa, D., Gritzalisb, D.: Long-term verifiability of the electronic healthcare records' authenticity. Int. J. Med. Informatics **76**, 442–448 (2007)
12. Lynch, C.: The future of personal digital archiving: defining the research agendas. In: From Personal Archiving: Preserving Our Digital Heritage, Information Today, vol. 50, May 2015
13. Estonian National Archive: The 2005–2010 National Archives' digital archives strategy, Tartu (2003)
14. Estonian National Archive: Digital archives vision, Tallinn (2005)

15. Estonian National Archive: Archives management requirements for digital records, Tallinn (2008)
16. Draheim, D., Koosapoeg, K., Lauk, M., Pappel, I., Pappel, I., Tepandi, J.: The design of the Estonian governmental document exchange classification framework. In: Kő, A., Francesconi, E. (eds.) EGOVIS 2016. LNCS, vol. 9831, pp. 33–47. Springer, Cham (2016). https://doi.org/10.1007/978-3-319-44159-7_3
17. Kalja, A., Robal, T., Vallner, U.: New generations of Estonian eGovernment components. In: Proceedings of PICMET 2015 – Portland International Conference on Management of the Technology Age, Portland (2015)
18. Atkinson, C., Draheim, D.: Cloud-aided software engineering: evolving viable software systems through a web of views. In: Mahmood, Z., Saeed, S. (eds.) Software Engineering Frameworks for the Cloud Computing Paradigm. CCN, pp. 255–281. Springer, London (2013). https://doi.org/10.1007/978-1-4471-5031-2_12
19. Draheim, D.: The service-oriented metaphor deciphered. J. Comput. Sci. Eng. **4**, 253–275 (2010)
20. Bordbar, B., Draheim, D., Horn, M., Schulz, I., Weber, G.: Integrated model-based software development, data access, and data migration. In: Briand, L., Williams, C. (eds.) MODELS 2005. LNCS, vol. 3713, pp. 382–396. Springer, Heidelberg (2005). https://doi.org/10.1007/11557432_28
21. Draheim, D., Nathschläger, C.: A context-oriented synchronization approach. In: Electronic Proceedings of the 2nd International Workshop in Personalized Access, Profile Management, and Context Awareness: Databases (PersDB 2008) in Conjunction with the 34th VLDB Conference (2008)
22. European Parliament and Council: Regulation (EU) No 910/2014 on Electronic Identification and Trust Services for Electronic Transactions in the Internal Market and Repealing Directive 1999/93/EC (eIDAS Regulation), European Union (2014)
23. Pappel, I., Pappel, I.: Methodology for measuring the digital capability of local governments. In: Proceedings of the 5th International Conference on Theory and Practice of Electronic Governance, Tallinn (2011)
24. Pappel, I., Pappel, I., Saarmann, M.: Development of information society and e-government by improving electronic records management solutions at Estonian local authorities. In: Kommers, P., Isaias, P. (eds.) Proceedings of the IADIS International Conference e-Society 2012, Berlin (2012)
25. Margulies, N.: Organizational Development: Values, Process, and Technology, p. 3. McGraw-Hill Book Co., New York (1972)
26. French, W.L., Bell, C.H.: Organization Development: Behavioral Science Interventions for Organization Improvement. Prentice-Hall, Englewood Cliffs (1998)
27. Calista, D.J., Melitski, J., Holzer, M., Manoharan, A.: Digitized government in worldwide municipalities between 2003 and 2007, pp. 588–600 (2010)
28. Pappel, I., Pappel, I., Saarmann, M.: Conception and activity directions for training and science centre supporting development of Estonian e-state technologies. In: Proceedings of the 5th International Conference on Theory and Practice of Electronic Governance, Tallinn (2011)
29. Pappel, I.: Paperless Management as a Foundation for the Application of e-Governance in Local Governments. Tallinn University of Technology, Tallinn (2014)
30. Hung, S.Y., Tang, K.Z., Chang, C.M., Ke, C.D.: User acceptance of intergovernmental services: an example of electronic document management system. Gov. Inf. Q. **26**(2), 387–397 (2009)

31. Yaacob, R.A., Mapong Sabai, R.: Electronic records management in Malaysia: a case study in one government agency. In: Asia-Pacific Conference on Library & Information Education & Practice 2011 (A-LIEP2011): Issues, Challenges and Opportunities, Malaysia (2011)
32. Sellen, A., Harper, R.: The Myth of the Paperless Office, pp. 17–18. MIT Press, Cambridge (2001)
33. Zachman, J.A.: Business systems planning and business information control study - a comparisment. IBM Syst. J. **21**(3), 31–53 (1982)
34. Winter, K., Buckl, S., Matthes, F., Schweda, C.M.: Investigating the state-of-the-art in enterprise architecture management methods in literature and practice. In: Sansonetti, A. (ed.) Proceedings of the 4th Mediterranean Conference on Information Systems, Tel Aviv (2010)
35. Taveter, K., Wagner, G.: A multi-perspective methodology for modelling inter-enterprise business processes. In: Arisawa, H., Kambayashi, Y., Kumar, V., Mayr, H.C., Hunt, I. (eds.) ER 2001. LNCS, vol. 2465, pp. 403–416. Springer, Heidelberg (2002). https://doi.org/10. 1007/3-540-46140-X_31
36. Draheim, D.: Smart business process management. In: 2011 BPM and Workflow Handbook, Digital Edition. Future Strategies, Workflow Management Coalition (2012)
37. Milani, F., García-Bañuelos, M., Dumas, M.: Blockchain and Business Process Improvement. BPTrends Newsletter (2016)
38. Norta, A., Ma, L., Duan, Y., Rull, A., Kõlvart, M., Taveter, K.: eContractual choreography-language properties towards cross-organizational business collaboration. J. Internet Serv. Appl. **6**, 1–23 (2015)
39. Wallace, C., Pordesch, U., Brandner, R.: Long-term archive service requirements (2007)
40. Levy, D., Schaettgen, N., Duvaud-Schelnast, J., Socol, S.: Digital signatures - paving the way to a digital Europe. Arthur D. Little (2014)
41. American City & Council: Benchmark Survey: Digital Signatures. ARX (2014)
42. Pappel, I., Pappel, I., Tampere, T., Draheim, D.: Implementation of e-invoicing principles in Estonian local governments. In: Proceedings of ECDG 2017 – the 17th European Conference on Digital Government, Lissabon (2017)
43. Sousa, P.R., Faria, P., Correia, M.E., Resende, J.S., Antunes, L.: Digital signatures workflows in alfresco. In: Kő, A., Francesconi, E. (eds.) EGOVIS 2016. LNCS, vol. 9831, pp. 304–318. Springer, Cham (2016). https://doi.org/10.1007/978-3-319-44159-7_22
44. Gorelik, S., Lyaper, V., Bershadskaya, L., Buccafurri, F.: Breaking the barriers of e-Participation: the experience of Russian digital office development. In: Kő, A., Francesconi, E. (eds.) EGOVIS 2014. LNCS, vol. 8650, pp. 173–186. Springer, Cham (2014). https://doi.org/10.1007/978-3-319-10178-1_14

Towards a Fine-Grained Privacy-Enabled Attribute-Based Access Control Mechanism

Que Nguyet Tran Thi$^{(\boxtimes)}$ and Tran Khanh Dang

Ho Chi Minh City University of Technology, Ho Chi Minh City, Vietnam
{ttqnguyet,khanh}@hcmut.edu.vn

Abstract. Due to the rapid development of large scale and big data systems, attribute-based access control (ABAC) model has inaugurated a new wave in the research field of access control. In this paper, we propose a novel and comprehensive mechanism for enforcing attribute-based security policies stored in JSON documents. We build a lightweight grammar for conditional expressions that are the combination of subject, resource, and environment attributes so that the policies are flexible, dynamic and fine grained. Besides, we also present an extension from the ABAC model for privacy protection with the approach of purpose usage. The notion of purpose is associated with levels of data disclosure and constraints to support more fine-grained privacy policies. A prototype built for the proposed model using Java and MongoDB has also presented in the paper. The experiment is carried out to illustrate the relationship between the processing time for access decision and the complexity of policies.

Keywords: Attribute based access control model
Purpose based access control model · Privacy protection · Privacy preserving

1 Introduction

Since the rapid development of large scale, open and dynamic systems, the shortcomings of traditional access control models (e.g. Discretionary Access Control (DAC), Mandatory Access Control (MAC), Role based Access Control (RBAC) [1]) have gradually revealed, for example, applied for only closed systems, role explosion, complexity in compulsory assignments between users, roles, and permissions, and inflexibility in specifying dynamic policies and contextual conditions. Attribute based access control models (ABAC) have been recently investigated [2–4] and considered as one of three mandatory features for future access control systems [5].

Extensible Access Control Markup Language (XACML) 3.0 is an industrial OASIS standard[1] for enforcing access control policies based on attributes, considered as a predecessor of ABAC. In XACML policies, every operation on attributes even trivial conditions such as comparison requires function and data type definitions. This has caused the verbosity and difficulty in the specification of policies. Moreover, XACML is based on XML, which is not well-suited for Web 2.0 applications. Meanwhile,

---

[1] https://www.oasis-open.org/committees/xacml/.

© Springer-Verlag GmbH Germany 2017
A. Hameurlain et al. (Eds.): TLDKS XXXVI, LNCS 10720, pp. 52–72, 2017.
https://doi.org/10.1007/978-3-662-56266-6_3

JavaScript Object Notation (JSON) language[2] is the fat-free alternative of XML. In [27], the experimental results indicate that JSON is remarkably faster and uses fewer resources than XML. Thus, JSON is currently a light weight and widely used data interchange format in the Web of Things. Moreover, since JSON is a subset of Java-Script, it is easier to parse components of a policy into programming objects for further processing. Besides, JSON has been used in many NoSQL databases for storage and retrieval with the high performance. Such advantages of JSON have brought the motivation for our work when using it to model attribute based policies.

Furthermore, our mechanism is built on the principle of the NIST Standard ABAC model that an access decision is *permitted* only if the request satisfies conditions on attributes of subject, resource and environment specified in policies. We also propose a light-weight grammar for conditional expressions, which are human readable text and enough robust to describe complex policies such as user, data, environment driven policies. Besides, we also build an additional module by extending the ABAC model for data privacy protection.

Privacy is a major concern in both of research and industrial fields due to dissemination of personal and sensitive data without user control, especially in mobile and ubiquitous computing applications and systems. In [7], privacy is defined as the claim of individuals, groups, or institutions to determine for themselves when, how, and to what extent information about them is communicated to others. Most previous studies have considered privacy protection in access control models as constraints on purpose of data usage. The research on purpose based access control (PBAC) model has recently drawn many interests, although it has developed since 2000s. However, to the best of our knowledge, no research has integrated PBAC into ABAC. The novel contribution of our work includes three main aspects: (1) using JSON to specify attribute based policies, (2) integrating PBAC model into ABAC model and (3) developing the prototype by Java and MongoDB database for demonstrating privacy preserving attribute based policy evaluation mechanism.

The rest of the paper is organized as follows. Section 2 gives a brief survey of related works. Section 3 presents the overview of our approach. In Sect. 4, we introduce the structure of policies and main components in our proposed model. Section 5 indicates the mechanism of the proposed access control model in details. The experiment for evaluating the processing time is shown in Sect. 6. Concluding remarks and future work are discussed in Sect. 7.

2 Related Work

The development of Information Technology, especially in the age of Big Data and Internet of Things, causes the role explosion problem and increases the complexity in permission management in RBAC models which have been dominant for a long time [20, 21]. An emerging interest in addressing these problems is ABAC models, which can be adaptable with large, open and dynamic environments [2, 3].

---

[2] www.json.org.

In the common approach of ABAC, according to the NIST standard [2], authorization decision is based on rules that simultaneously specify a set of conditions on numerous attributes such as subject, object, action and environment for a certain valid permission. There are many research works on ABAC. In [28], the authors have presented a taxonomy of ABAC research which is demonstrated in Fig. 1. According to the classification, our work focuses on the branches as follows: *ABAC models, policy language* and *confidentiality of attributes*. In this survey, ABAC research about confidentiality of attributes means that how to ensure the privacy of attributes in the model. It has been also recognized that no research work in the survey have used JSON for policy language and addressed data privacy protection in ABAC.

There are two fields for researching about ABAC models, pure ABAC models and hybrid models. Several papers have provided various approaches for a general model. In [4], the authors took a first consideration about formal connections between traditional access control models (DAC, MAC, RBAC) and ABACα which consists of users, subjects, objects and their attributes. In this ABACα, the authors did not mention the environment component in policies as well as the enforcement mechanisms. In [2, 3], the authors provided a definition of ABAC and considered about using ABAC in organizations according to NIST standard. However, the implementation has not been discussed yet in these papers. In another paper [29], Next Generation Access Control (NGAC) takes advantages of graphs illustrating assignment and association between attributes and values to perform access rights. It has provided benefits for policy review and management. However, the cost for building such graphs and the complexity of the policy evaluation algorithm increase significantly when the size of attribute domains and the variety of data structure grow up. In hybrid models, majority of research works have integrated traditional access control models (DAC, MAC, and RBAC) with ABAC such as [21–23]. About *policy language* to express policy enforcement mechanism in the above papers, several approaches such as XACML, logic programming languages or UML have been proposed. For the last problem, the *confidentiality of attributes*, access control models need to provide a mechanism which can protect data attributes with the highest fine-grained level as possible. Attribute based access control models can allow or not allow to access to each data attribute in various context through attribute based policies. Purpose based access control (PBAC) model is another approach to protect data privacy based on the concept of purpose of data usage. A purpose compliance check in PBAC models depends on the relationship between access purposes and intended purposes of data objects ranging from the level of tables to the data cells [7–10]. In the beginning, Byun et al. [7] proposed the model with two types of allowable and prohibited intended purposes. It was then extended with an additional purpose, i.e. conditional intended purpose [9]. Several works have been conducted on enhancing this model by combining with role based access control (e.g., [11–14]), implementing with relational database management systems (DBMSs) with the technique of SQL query rewriting [15] and integrating with MongoDB [16]. Recently, action-aware with indirect access and direct access has also been considered in policies [17]. Nevertheless, research works about PBAC have not expressed privacy policies with the approach of attribute based policies yet except of the paper in [30]. However, such PBAC models have not provided the fine granularity for privacy policies. In such approaches, there are three setting levels for disclosing values of

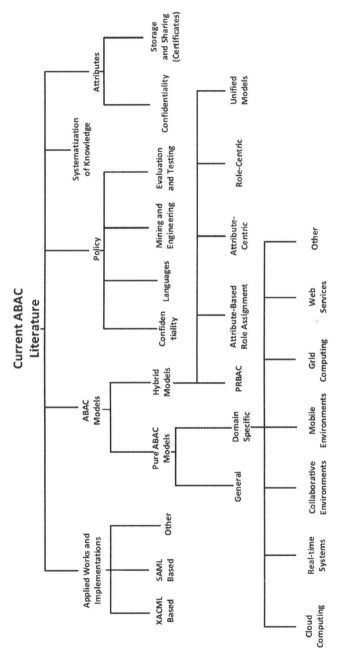

Fig. 1. A taxonomy of attribute based access control research [28]

attributes, that is, *show, hide* and *partial*. All data items of the same attribute have the same disclosure level for the same purpose. Besides, such privacy policies have also not considered about contextual conditions.

With the different approach compared to related works, our research takes a new mechanism for enforcing attribute based policies using the JSON language and proposes an additional module for data privacy protection based on the principle of PBAC and its enhancement.

In summary, our work contributes a novel and comprehensive attribute based access control mechanism which can preserve data privacy. In this mechanism, we use JSON which is more light-weight than XML and widely used in Web 2.0 applications as the language for policy specification. The additional module, *PurJABAC*, is integrated into the model based on the purpose concept to describe attribute based data privacy policies. Our approach can support fine grained policies with contextual and attribute based constraints and protect data attributes with various disclosure levels according to the tree of data generalization.

3 The Overview

In this paper, we propose a model, called as the JABAC model which is an integration between ABAC, PBAC and using JSON to express attribute based security policies to regulate data accesses and protect data privacy. We also provide a mechanism to execute this model. In this section, we briefly describe the access control mechanism of our proposed model. In Fig. 2, requests from applications are sent to the *Policy Enforcement Point* module. For each request, all necessary data is retrieved. Both of data and the request are processed and converted to the *JSON based request context*. After that, it is sent to the *JABAC* module to be decided whether it is permitted or denied according to the predefined attribute based policies stored in *JSON document store database*. In addition, before returning data to the requester, *JABAC* calls the *PurJABAC* module to filter data according to privacy policies. We achieve privacy awareness through the *PurJABAC* module which enforces privacy policies to show, hide and generalize data before the requester receives them.

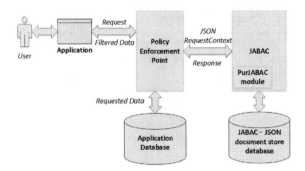

Fig. 2. Overview of the approach

A *JSON based request context* contains information of a request string and the context at the time of requesting. Any request issued from the application is converted to the structure of a *request context* in the JSON format by *Policy Enforcement Point*. If there exists at least one rule requiring requested data during the process of policy evaluation, the data queried from the application database will be converted into the JSON format and filled in the *request context* for processing.

After evaluating necessary policies, *JABAC* returns a response to *Policy Enforcement Point* for the decision result. The response also contains the final data if they are filtered according to related privacy policies by the *PurJABAC* module. *Policy Enforcement Point* has the responsibility to convert data into the format of application.

The communication between the application, the application database and *JABAC* uses the JSON format to exchange data. Therefore, *JABAC* is independent from the technology of application and application database.

Unlike the traditional purpose based access control models, in the approach, all degrees of privacy policies from the table one to the cell one are expressed in the same way under the form of attribute based policies through generalization functions. Furthermore, we take attribute based conditions into consideration for privacy policies.

The structure of access control and privacy policies is identical. Therefore, in general, we provide a simple but sound and comprehensive solution. The details of the JABAC model and its mechanism will be provided in Sects. 4 and 5.

4   The JABAC Model

We describe the proposed access control model in this section before describing the mechanism to enforce this model. When a subject s accesses an object o, the authorization process is carried out through two stages called as 2-stage authorization: (1) access policy authorization and (2) privacy policy authorization. The first step using access policies verifies that the request is legitimate with rights for the subject to access data. After that, the request is transferred to the second stage for checking privacy compliance based on privacy policies.

Access policies describe access rights of subjects on resources, and conditions compositing of attributes of subject, action, resource, and environment as well as obligations that are instructions from *Policy Decision Point* to *Policy Enforcement Point* to be performed before or after data results is returned to the requester.

Privacy policies describe access restriction on data objects which need to be preserved privacy. The structure of privacy policies is similar to access policies. It also includes subject, action, resource, environment, obligations and attribute based condition. However, each component contains the slightly different meaning. The components of subject, action, resource, environment in privacy policies indicate attribute based conditions and the component of obligation contains *generalization functions* applied on values of data objects for privacy protection.

In privacy preserving access control, the purpose concept plays an important role in privacy policies to describe a valid reason for data usage. When users send a request to query data, they must provide their access purpose to the system. The access purpose is then verified whether the subject is permitted for using it in the access policy authorization stage. In our model, this value, access purpose, is considered as an attribute of the environment entity. However intended purpose is not implicitly mentioned, it is described through a conditional expression based on the access purpose attribute identifying which values are valid for data usage and generalization (Fig. 3).

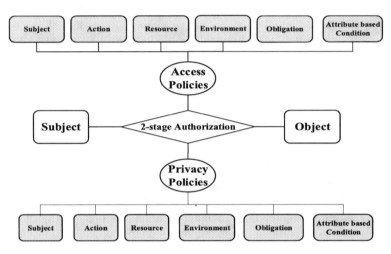

Fig. 3. The components of ABAC model for privacy protection

4.1   Policy Structure

In this section, we present a general structure used for both of access and privacy policies. However, for each use case, we will describe how to specify policies through examples. The mechanism of processing policies works slightly differently.

General Structure. In our model, a *policy set* includes *policies*. Each *policy* includes *rules*. Each *rule* defines a conditional expression that is a critical component in the policy. The rule returns a value specified in *Effect* if the condition is true. The *target* component, including three sub components *Subject*, *Action* and *Resource*, is used to pre-select applicable policies for access decision. To avoid conflicts between policies and rules, the policy and rule combining algorithms such as *permit override*, *deny override*, etc. are applied into the policy set and policies. The implementation for rule combining algorithm is inherited from XACML [26]. The final component in a rule is *obligations* indicating actions which will be performed before or after a final response is established by *Policy Enforcement Point*. The relationship diagram between policy set, policies and rules are illustrated in Fig. 4. The structure of a policy in the JSON format is illustrated in Fig. 5. Its properties are described in details in Table 1.

In the above structure, the *condition* component *<conditionalExpr>* is written according to the below grammar (Fig. 6):

It can be seen that the operands in the condition expression are attributes from *Subject, Action, Resource, ResourceContent* and *Environment* or specific values. The values of attributes are loaded from the *request context*. For missing values, *JABAC*

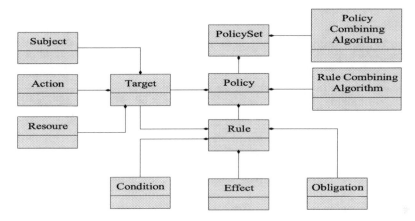

Fig. 4. The relationship diagram of components in policy set

```
PolicySet {
  PolicyCombiningAlgID: <CombiningPolicyAlgID>,
  Policies: [{
      PolicyID: <policyID>,
        RuleCombiningAlgID: <CcombiningRuleAlgID>,
      Target: {
        Subject: <subjectID>,
        Action: <actionID>,
        Resource: <resourceName>},
      Rules: [{
        RuleID: <ruleID>,
        Target: {
          Subject: <subjectID>,
          Action: <actionID>,
          Resource: <resourceName>},
        Condition: <conditionalExpr>,
        Effect: <"permit" or "deny">,
        Obligations: [{
          FunctionID: <functionName>,
          Parameters: [{
            ParaValue: <attributeName>|<specificValue>,
            SourceType: <entityName>|Null,
            DataType: <dataType>}],
          FullFillOn: <"permit" or "deny">,
          Directive: <"before" or "after">}]
        }]}]}
```

Fig. 5. The JSON structure of policy

Table 1. The description of the fields in a policy

Level	Field name	Type	Description
PolicySet	PolicyCombiningAlgID	String	Indicate which combining algorithm is used to combine the results of evaluating policies in the policyset
PolicySet. Policy	Policies	Array of documents	Contain a list of policies
PolicySet. Policy	PolicyID	String	Indicate policy code
PolicySet. Policy	RuleCombiningAlgID	String	Indicate which combining algorithm is used to combine results of rule evaluation in a policy
PolicySet. Policy	Target	Embedded document	Indicate which subject, action, resource is applied for the policy
PolicySet. Policy.Target	Subject	String	Indicate ID of subject or ANY
PolicySet. Policy.Target	Action	String	Indicate ID of action or ANY
PolicySet. Policy.Target	Resource	String	Indicate name of resource or ANY
PolicySet. Policy	Rules	Array of embedded documents	Contain a list of rules in each policy
PolicySet. Policy.Rules	RuleID	String	Indicate rule code
PolicySet. Policy.Rules	Target	Embedded document	Indicate which subject, action, resource is applied for a rule. Similar to PolicySet.Policy. Target
PolicySet. Policy.Rules	Condition	String	Contain a Boolean expression. If the condition is evaluated as true, the rule returns the value of the effect property
PolicySet. Policy.Rule	Effect	String	Indicate whether the rule returns permit or deny if the condition is evaluated as true
PolicySet. Policy.Rule	Obligations	Array of embedded documents	Contain a list about obligations which will be executed after evaluating policies
PolicySet. Policy.Rule. Obligations	FunctionID	String	Contain the function name which will be called to execute for an obligation

(continued)

Table 1. (*continued*)

Level	Field name	Type	Description
PolicySet. Policy.Rule. Obligations	Parameters	Array of embedded documents	Contain a list of parameters of the function defined in FunctionID, including ParaValue, SourceType, and ParaType
PolicySet. Policy.Rule. Obligations. Parameters	ParaValue	Object	Contain the value for the parameter
PolicySet. Policy.Rule. Obligations. Parameters	SourceType	String	If the ParaValue is a certain attribute name, SourceType contains the name of source of attribute to get value for that attribute. Example, ParaValue is "subjectName" and SourceType is "Subject". If SourceType has Null value, the ParaValue field contains a specific and direct value
PolicySet. Policy.Rule. Obligations. Parameters	ParaType	String	The name of data type of ParaValue. This is used to convert into the data type of the corresponding argument declared in the function
PolicySet. Policy.Rule. Obligations	FullFillOn	String	Contain the value of *permit* or *deny* which indicates the case of executing an obligation. The *Permit* value means that the obligation will be executed when the final result is Permit and vice versa for the Deny value
PolicySet. Policy.Rule. Obligations	Directive	String	Indicate when the obligation is executed, namely, before or after data results are returned to the subject

will look up from database to fulfill the request context. More details will be presented in Sect. 5.

A below example for a rule demonstrates the policy structure and the grammar of conditional expression:

Doctors can read their patient records with the purpose of treatment. If the request is denied, the system will email to the administrator John about the subjectID and the access purpose of subject.

```
RuleID: "ARU001",
Target: {
    Subject: "ANY",
    Action: "read",
    Resource: "patients"},
Condition: "Subject.role = 'doctor' AND
        ResourceContent.doctorID = Subject.subjectID
        AND Environment.accessPurpose = 'treatment'",
Effect: permit,
Obligations:[{
    FunctionID: "email",
    Parameters: [{
        ParaValue: "john@gmail.com",
        SourceType: Null,
        DataType: "String"},
        {ParaValue: "subjectID",
        SourceType: "Subject",
        DataType: "String"}],
    FullFillOn: "deny",
    Directive: "after"}]
```

With the above rule, a doctor can read only patient records if he is a doctor and his *subjectID* equals to the *doctorID* in each patient record. By obligations, if the request is denied, the information of *subjectID* will be sent to *john@gmail.com* by calling the function *email* after the response is returned to the requester. The attributes *role, subjectID* of subject, the attribute *accessPurpose* of environment, and the attribute *doctorID* of resource content keep values in the *request context*. If they cannot be found, the system will call *Policy Information Point* to look up them in the *application database* and the *JABAC database*. If an error occurs or some values are missing, the rule will return *Indeterminant*. If the target component does not match with the information in the request context, the rule will return *NotApplicable*. To produce the final result, the mechanism will take advantage of the rule and policy combing algorithms specified in the policy to make a decision when evaluating rules and policies in the loop. The structure of *request context* and details of the mechanism will be presented in Sects. 4.2 and 5 respectively.

When a request is evaluated as permit, it does not ensure that all data corresponding to the request will be accessed by the subject. For example, Alice with the role of doctor can read the records of her patients according to the above rule. However, due to privacy, Bob, one of her patients, only wants his information about address, social security number, and birthdate to be partially shown to his doctor. Thus, in our proposed model, Bob can define *privacy policies* to protect his data. However, it can appear special cases defined by the highest security administrative to bypass his privacy policies such as standard regulations of the organization. In this paper, such delegation problem has not been mentioned yet.

```
grammar ConditionalExpression;

condition          : logical_expr EOF;
logical_expr       : logical_expr AND  logical_expr       # LogicalExpressionAnd
                   | logical_expr OR logical_expr         # LogicalExpressionOr
                   | LPAREN logical_expr RPAREN           # LogicalExpressionInParen
                   | comparison_expr                      # ComparisonExpression
                   | BOOLEAN                               #| LogicalEntity
                   ;

comparison_expr : comparison_operand
                  atomic_compare comparison_operand       # ComparisonAtomicCompare
                | comparison_operand
                  set_compare comparison_operand          # ComparisonSetCompare
                   ;

comparison_operand : arithmetic_expr;

atomic_compare   : GT | GE | LT | LE | EQ | NE
                   ;
set_compare      : 'IN' | 'EQ';

arithmetic_expr  : MINUS arithmetic_expr               # ArithmeticExpressionNegation
                 | LPAREN arithmetic_expr RPAREN       # ArithmeticExpressionParens
                 | operand                             # ArithmeticExpressionDataEntity
                 | arithmetic_expr MULT arithmetic_expr # ArithmeticExpressionMult
                 | arithmetic_expr DIV arithmetic_expr  # ArithmeticExpressionDiv
                 | arithmetic_expr PLUS arithmetic_expr # ArithmeticExpressionPlus
                 | arithmetic_expr MINUS arithmetic_expr # ArithmeticExpressionMinus
                   ;

operand          : 'Subject.' ID                       # OperandSubjectAttribute
                 | 'Action.' ID                        # OperandActionAttribute
                 | 'Resource.' ID                      # OperandResourceAttribute
                 | 'Environment.' ID                   # OperandEnvironmentAttribute
                 | 'ResourceContent.' field_name       # OperandResourceContent
                 | constant                            # OperandConstant
                   ;

constant         : DECIMAL                             # ConstantNumber
                 | STRING                              # ConstantString
                 | '[' constant (',' constant)* ']'    # OperandArrayConstant
                 | '[' ']'                             # OperandArrayEmpty
                   ;

field_name       : (ID ('[' INDEX ']')?)+;
filter_operation: '$eq' | '$gt' | '$gte' | '$lt' | '$lte' | '$ne' | '$in' | '$nin';
AND : 'AND'; OR  : 'OR' ;

ID                 : [a-zA-Z_][a-zA-Z_0-9]+ ;
INDEX              : '.'[0-9]+;
FIELD              : '.'[a-zA-Z_][a-zA-Z_0-9]+;
DECIMAL            : '-'?[0-9]+('.'[0-9]+)? ;
STRING             : '\'' (~('\\'|'\''))* '\'';
BOOLEAN            : 'true'|    'false';
WS                 : (' '|'\t')+ {skip();} ;
```

Fig. 6. The grammar of conditional expression

Privacy Policies. They have the same structure with the general one but it is slightly different about the use of components. The obligations in privacy policy play the role of expressing *how to generalize* the value of data item to protect privacy. In our mechanism, an obligation in privacy policies associates with the special function *MakeGeneralization (field name, data disclosure level)* for hiding details of data.

It takes two parameters; *the name of the field* which needs to be protected and *the level of data disclosure* for privacy preserving. The definition of *data disclosure level* is presented as below.

Data Disclosure Level (DL). DL of data item represents the level of data generalized in the *Domain Generalization Hierarchy* (DGH). Based on DL, data are generalized into a new value according to *Value Generalization Hierarchy* (VGH) generated from DGH. The concepts of VGH and DGH are explained as follows: Each attribute has a range of values designated by a domain. For data privacy preserving, the generalization process is applied to the value domain of the attribute and establishes a *DGH tree*. Each value in the domain contains many generalized value in generalized domains. A set of these values with the order of generalization process establishes a *VGH tree*. The formal definitions for DGH and VGH tree can be seen in [24].

Figure 7 describes a *DGH tree* and *VGH tree* for the attribute domain *Birthdate*. In this example, the number of data disclosure levels of birthdate is five, in which the smallest number (DL0) indicates the data does not require any privacy protection, whereas the highest number indicates the data will be hidden with the keyword "*".

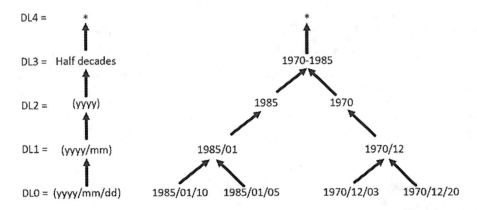

Fig. 7. Domain generalization hierarchy tree

In summary, to specify policies for privacy protection, *access purpose* is defined as an attribute of environment; obligations is modeled as an indicator for data generalizations. For example, the below privacy rule PRU001 indicates "The information about birthdate and social security number of Bob who has the *patientID* 10001 are generalized according to the following levels: 2 (only displaying the year of birthdate), and 1 (not displaying any information of the string) respectively when any subject read the patient record of Bob with the access purpose of treatment". Thus, the rule PRU001 is specified as follows:

The components such as *Effect* and *FullFillOn* will be discarded because they have no meaning in privacy policies. The *PurJABAC* module retrieves privacy policies and applies the *MakeGeneralization* function for related fields in each document in ResourceContent.

```
RuleID: "PRU001",
Target: {
    Subject: ANY,
    Action: read,
    Resource: patients},
Obligations: [
    {FunctionID: "MakeGeneralization",
    Parameters: [
        {ParaValue: "Birthdate",
        SourceType: "ResourceContent",
        DataType: "String"},
        {ParaValue: 2,
        DataType: "Integer"}],
    FullFillOn:  Null},
    {FunctionID: "MakeGeneralization",
    Parameters: [
        {ParaValue: "SSN",
        SourceType: "ResourceContent",
        DataType: "String"},
        {ParaValue: 1,
        SourceType: Null,
        DataType: "Integer"}],
    FullFillOn:  Null}],
Condition: "ResourceContent.patientID = '10001' AND
Environment.accessPurpose = 'treatment'",
Effect: null
```

Take an example that Alice, who has the subjectID *20001*, sends a request to database to read her patient records with the access purpose "treatment". Alice is allowed but information about birthdate and social security number of Bob is generalized by PRU001. For instance, Table 2 shows patient records in the database. There are only two records of Bob and Kitty which Alice can see due to the rule *ARU001*. Besides, the information of Bob displays only the year of his birthdate and his ssn with the special character "*" due to the rule *PRU001* (ref. Table 3). In the next section, the structure of the request context is presented in details.

Table 2. An example of patient records

patientID	patientName	birthdate	ssn	doctorID
10001	Bob	1/13/1990	12345789	20001
10002	Paul	12/10/1980	12345999	20002
10003	Kitty	3/13/1970	12345777	20001

Table 3. An example of patient records returned to alice

patientID	patientName	birthdate	ssn	doctorID
10001	Bob	*1990*	*	20001
10003	Kitty	3/13/1970	12345777	20001

4.2   The Request Context Structure

The structure of a *request context* is very important for the *JAPAC* mechanism. A *request context* contains the components such as *Subject, Resource, Action* and *Environment*. The below example demonstrates the structure of the *request context*. Each

```
Subject: {
    SubjectID: "Alice",
    Attributes: [{
                AttributeID: "Role",
                AttributeType: "String",
                Value: "doctor"}]},
Resource: {
    ResourceID: "db.Patients",
    ResourceRequest: "db.Patients.find({doctorID:
                      '20001'})",
    ResourceContent: [
        { patientID : "10001",
          patientName: "Bob",
          birthDate:"1/13/1990",
          ssn:"12345789",
          doctorID:"20001"},
        { patientID : "10003",
          patientName: "Kitty",
          birthDate:"3/13/1970",
          ssn:"12345777",
          doctorID:"20001"}],
    Attributes: [{
                AttributeID: "SelectedFields",
                AttributeType: "Array",
                Value: "['patientID',
    'patientName', 'birthDate' 'ssn', 'doctorID']"}]},
Action: {
    ActionID: "READ",
    Attributes: []}
Environment: {
    EnvironmentID: Null,
    Attributes: [{
                AttributeID: "WorkingTime",
                AttributeType: "DateTime",
                Value: "14:30"}]}
```

part contains ID and its attributes. Especially, the *Resource* component includes *ResourceContent* which contains data queried from the original request and inserted by *Policy Enforcement Point* into the request context.

In our model, *JABAC* receives a request which must comply with the designed structure in the JSON format and contains information about the request such as subject, action, resource and environment. The system then analyzes the request and evaluates whether it is permitted or denied. After evaluating, *JABAC* calls the *PurJABAC* model to filter data for privacy protection.

In this section, the structure of components has been introduced in details. However, the mechanism for enforcing such policies has not been mentioned yet. In the next section, we present the mechanism of evaluating access control policies and processing data based on privacy policies.

5   The JABAC Mechanism

In this work, we utilize the concepts in XACML such as *Policy Enforcement Point* (PEP), *Policy Decision Point* (PDP), *Policy Information Point* (PIP), *Policy Administration Point* (PAP) and *Obligations*. PEP takes responsibility for receiving requests from applications, converting into request contexts in the JSON format, sending them to PDP for evaluating whether they are permitted or not, filtering and hiding data according to privacy policies and finally executing obligations of related access control policies. PDP receives the JSON request contexts to check access rights with access control policies. PIP has the feature of collecting information of attributes if PDP cannot find them in the request contexts. PAP represents the module of specifying policies. Obligation is the module of executing obligation functions of access control policies. With obligations specified in privacy policy, the PurJABAC module has a different mechanism to enforce. The below section will describe the data flow of our model.

The main processes in the data flow of our model depicted in Fig. 8 are described as follows:

1. **PEP** receives the access request consisting of the components: *subject, action,* and *resource.*
2. **PEP** creates another request, called *request context,* for policy decision from the access request fulfilled with the attributes of subject, action, resource, and environment and then sends to **PDP** for access authorization.
3. **PDP** retrieves the list of access policies from database.
4. For each policy, **PDP** checks whether the *target* element of the policy (i.e. *subject, action, resource*) matches with the corresponding components of the request context by the *Target Matching* module. If it returns "*successfully matching*", all rules of this policy are examined. Rules satisfying the *target* component will be processed in the next step.

5. In this step, the *condition* component of applicable rules is evaluated by the module **ConditionalExpr Parser and Evaluator**. **ConditionalExpr Parser and Evaluator** uses the open source *ANTLR*[3] with the grammar presented in Sect. 4.1 to evaluate the expression.

6. The module **Condition Parser and Evaluation** sends requests to *Policy Information Point* (**PIP**) to retrieve values for operands.

7. **PIP** collects values from the request context and database; and then sends them to **ConditionalExpr Parser and Evaluator** for expression evaluation.

   Depending on the combining rule algorithm specified in the component PolicyCombiningAlgID in the current policy, PDP will continue to check the next access rule (e.g. permit overrides) or terminate with the result deny (e.g. deny overrides). Similarly, depending on the combining algorithms for policies specified in the component RuleCombiningAlgID of PolicySet, PDP will terminate together with the result of policy evaluation or keep on checking with other applicable policies.

8. After evaluating all applicable policies from step 4 to step 7, **PDP** returns the response to **PEP**. The response can be *permit* or *deny*.

9. In the case of permit, **PEP** asks the **PurJABAC** module for filtering and hiding data in *ResourceContent* of the request context.

   To generalize data for privacy protection, **PurJABAC** retrieves privacy policies which their *target* component matches with the request context and then read data documents in *ResourceContent*. For each document, **PurJABAC** selects the set of privacy policies which can be applied to the current document and then computes the highest disclosure level for each field name from those privacy policies. Take the example in Sect. 4; we assume that there have two privacy rules for Bob and one privacy rule for Kitty. The first one is PRU001 described above and the second one is PRU002 which allows showing the year and the month of birthdate (the lower level than PRU001) in the case of treatment purpose. Thus, PRU001 wants to generalize the birthdate of Bob into the year level meanwhile the birthdate of Kitty is not influented by PRU001. Therefore, the **PurJABAC** module chooses the highest disclosure value, DL02 for Bob and DL01 for Kitty to generalize their birthdate.

10. In the case *ResourceContent* has not been fulfilled in the request context due to no prior policy requires resource data, the **PurJABAC** module will send the request to **PIP** for querying resource content.

11. After checking all objects in the result set, PEP calls *Obligation Services* to perform obligations. The obligations are executed before or after PEP returns results to the requester. If any function in obligations returns an exception, PEP will not send data to the requester however the access is permitted. For example, an obligation requires the requester to accept terms and conditions. If she/he refuses to perform this obligation, her request is denied.

12. **PEP** returns data results to the requester.

---

[3] www.antlr.org.

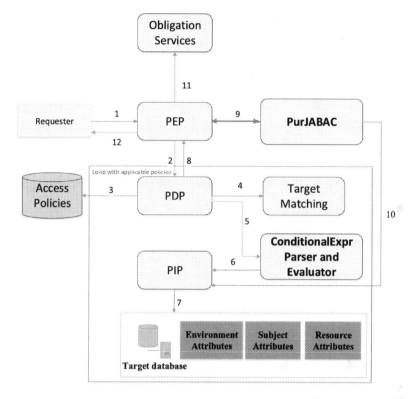

Fig. 8. The data flow diagram of the proposed access control model

In general, the components of our model and its mechanism are presented in this paper. With our best knowledge, no research work has integrated PBAC into ABAC. Besides, we extend the fine grained feature for access control policies through the conditional expression specified by our grammar and for privacy policies through the generalization function with various data disclosure levels.

6   Evaluation

We carried out experiments about the relevance between processing time of the PDP module and complexity of rules. The system configuration for the experiments is Dell Vostro 3650, 8 GB RAM, Intel core i5-3230M 2.60 GHz. The prototype is implemented by Java SDK, Spring Framework and MongoDB 3.0.7 for storing policies and data. The target database includes 20 collections with 20 attributes and 200 documents for each collection generated randomly. Each subject, action, resource and environment contains 20 attributes.

The following Table 4 indicates the results after five experiments. For each experiment, we measure processing time with three times (e.g. T1, T2 and T3). From the Table 1, it can be seen that the processing time increases with the complexity of

Table 4. The Results of Experiments

ID	Number of rules	Logical expressions in each rule	Arithmetic expressions in each rule	T1 (ms)	T2 (ms)	T3 (ms)
1	1	1	1	5	4	4
2	1	10	10	8	8	7
3	10	10	10	12	11	12
4	50	10	10	15	14	14
5	100	10	10	17	17	16

rules. However, when there was a fivefold and twofold increase in the number of rules from 10 to 50 and from 50 to 100, the processing time only increased by 2–3 ms, approximately 13–25%.

About the fine granularity, our attribute based access control model can describe various attribute based policies due to the flexibility of conditional expressions built under the proposed grammar. Compared to other approaches, our model can specify policies with the conditions based on the combination of attributes of subject, resource, and environment. Besides, by using JSON, our model can easily integrate with Web 2.0 applications as well as devices and systems in Internet of Things.

7   Conclusion and Future Work

In this paper, we have proposed the fine grained attribute and purpose based access control model *JABAC* for privacy protection with the mechanism of 2-stage authorization. A conditional expression based on attributes of subject, resource, action and environment are built on the ANTLR grammar, which is enough to describe various policies. Besides, an additional module *PurJABAC* is integrated into JABAC for privacy protection. In our approach, privacy policies are specified under the same structure of attribute based policies. Purposes are associated with the new concept, disclosure level, for indicating the degree of generalization for privacy protection. They are demonstrated as obligations in privacy policies which call *MakeGeneralization* functions to generalize data.

However our model takes a novel approach in the research field of attribute access control and purpose based access control, there have still some disadvantages. In future, we will improve the grammar for conditional expressions to describe more complex policies which can retrieve data from multiple sources not only from the request context. Furthermore, the solution currently supports the data which are not allowed many levels of embedded documents. Therefore, the complexity of data will be investigated as well. Besides, the problems such as policy review and administration have not been mentioned yet in this article. They will be promising research problems in the future.

Acknowledgements. This research is funded by Vietnam National University Ho Chi Minh City (VNU-HCM) under grant number C2017-20-11.

References

1. Bertino, E., Ghinita, G., Kamra, A.: Access Control for Databases: Concepts and Systems. Now Publishers, Hanover (2011)
2. Hu, V.C., Ferraiolo, D., Kuhn, R., Friedman, A.R., Lang, A.J., Cogdell, M.M., Schnitzer, A., Sandlin, K., Miller, R., Scarfone, K.: Guide to Attribute Based Access Control (ABAC) definition and considerations (draft). NIST Special Publication, 800, 162 (2013)
3. Hu, V.C., Kuhn, D.R., Ferraiolo, D.F.: Attribute-based access control. Computer 2, 85–88 (2015)
4. Jin, X., Krishnan, R., Sandhu, R.: A Unified attribute-based access control model covering DAC, MAC and RBAC. In: Cuppens-Boulahia, N., Cuppens, F., Garcia-Alfaro, J. (eds.) DBSec 2012. LNCS, vol. 7371, pp. 41–55. Springer, Heidelberg (2012). https://doi.org/10.1007/978-3-642-31540-4_4
5. Sandhu, R.: The Future of access control: attributes, automation, and adaptation. In: Krishnan, G.S.S., Anitha, R., Lekshmi, R.S., Kumar, M.S., Bonato, A., Graña, M. (eds.) Computational Intelligence, Cyber Security and Computational Models. AISC, vol. 246, p. 45. Springer, New Delhi (2014). https://doi.org/10.1007/978-81-322-1680-3_5
6. Westin, A.F.: Privacy and Freedom. Atheneum, New York (1967)
7. Byun, J.-W., Bertino, E., Li, N.: Purpose based access control of complex data for privacy protection. In: Proceedings of the Tenth ACM Symposium on Access Control Models and Technologies (2005)
8. Byun, J.W., Li, N.: Purpose based access control for privacy protection in relational database systems. VLDB J. 17(4), 603–619 (2008)
9. Kabir, M.E., Wang, H.: Conditional purpose based access control model for privacy protection. In: Proceedings of the Twentieth Australasian Conference on Australasian Database, vol. 92, pp. 135–142. Australian Computer Society, Inc. (2009)
10. Wang, H., Sun, L., Bertino, E.: Building access control policy model for privacy preserving and testing policy conflicting problems. J. Comput. Syst. Sci. 80(8), 1493–1503 (2014)
11. Kabir, M.E., Wang, H., Bertino, E.: A role-involved conditional purpose-based access control model. In: Janssen, M., Lamersdorf, W., Pries-Heje, J., Rosemann, M. (eds.) EGES/GISP 2010. IFIP AICT, vol. 334, pp. 167–180. Springer, Heidelberg (2010). https://doi.org/10.1007/978-3-642-15346-4_13
12. Kabir, M.E., Wang, H., Bertino, E.: A conditional purpose-based access control model with dynamic roles. Expert Syst. Appl. 38(3), 1482–1489 (2011)
13. Ni, Q., Lin, D., Bertino, E., Lobo, J.: Conditional privacy-aware role based access control. In: Biskup, J., López, J. (eds.) ESORICS 2007. LNCS, vol. 4734, pp. 72–89. Springer, Heidelberg (2007). https://doi.org/10.1007/978-3-540-74835-9_6
14. Ni, Q., Bertino, E., Lobo, J., Brodie, C., Karat, C.M., Karat, J., Trombeta, A.: Privacy-aware role-based access control. ACM Trans. Inf. Syst. Secur. (TISSEC) 13(3), 24 (2010)
15. Colombo, P., Ferrari, E.: Enforcement of purpose based access control within relational database management systems. IEEE Trans. Knowl. Data Eng. 26(11), 2703–2716 (2014)
16. Colombo, P., Ferrari, E.: Enhancing MongoDB with purpose based access control. IEEE Trans. Depend. Secur. Comput. (2015, will appear)
17. Colombo, P., Ferrari, E.: Efficient enforcement of action-aware purpose-based access control within relational database management systems. IEEE Trans. Knowl. Data Eng. 27(8), 2134–2147 (2015)
18. Pervaiz, Z., Aref, W.G., Ghafoor, A., Prabhu, N.: Accuracy-constrained privacy-preserving access control mechanism for relational data. IEEE Trans. Knowl. Data Eng. 26(4), 795–807 (2014)

19. Ferraiolo, D.F., Sandhu, R., Gavrila, S., Kuhn, D.R., Chandramouli, R.: Proposed NIST standard for role-based access control. ACM Trans. Inf. Syst. Secur. (TISSEC) **4**(3), 224–274 (2001)

20. Fuchs, L., Pernul, G., Sandhu, R.: Roles in information security–a survey and classification of the research area. Comput. Secur. **30**(8), 748–769 (2011)

21. Kuhn, D.R., Coyne, E.J., Weil, T.R.: Adding attributes to role-based access control. IEEE Comput. **43**(6), 79–81 (2010)

22. Huang, J., Nicol, D.M., Bobba, R., Huh, J.H.: A framework integrating attribute-based policies into role-based access control. In: Proceedings of the 17th ACM symposium on Access Control Models and Technologies, pp. 187–196. ACM (2012)

23. Rajpoot, Q.M., Jensen, C.D., Krishnan, R.: Attributes enhanced role-based access control model. In: Fischer-Hübner, S., Lambrinoudakis, C., Lopez, J. (eds.) TrustBus 2015. LNCS, vol. 9264, pp. 3–17. Springer, Cham (2015). https://doi.org/10.1007/978-3-319-22906-5_1

24. Sweeney, L.: Achieving k-anonymity privacy protection using generalization and suppression. Int. J. Uncertain. Fuzziness Knowl. Based Syst. **10**(05), 571–588 (2002)

25. Ni, Q., Bertino, E., Lobo, J.: An obligation model bridging access control policies and privacy policies. In: Proceedings of the 13th ACM Symposium on Access Control Models and Technologies, pp. 133–142 (2008)

26. Rissanen, E.: eXtensible Access Control Markup Language (XACML) version 3.0 (committe specification 01). Technical report, OASIS (2010). http://docs.oasisopen.org/xacml/3.0/xacml-3.0-core-spec-cd-03-en.Pdf

27. Nurseitov, N., et al.: Comparison of JSON and XML data interchange formats: a case study. In: Caine 2009 (2009)

28. Servos, D., Osborn, S.L.: Current research and open problems in attribute-based access control. ACM Comput. Surv. (CSUR) **49**(4) (2017)

29. Ferraiolo, D., et al.: Extensible Access Control Markup Language (XACML) and Next Generation Access Control (NGAC). In: Proceedings of the 2016 ACM International Workshop on Attribute Based Access Control (2016)

30. Thi, Q.N.T., Si, T.T., Dang, T.K.: Fine grained attribute based access control model for privacy protection. In: Dang, T.K., Wagner, R., Küng, J., Thoai, N., Takizawa, M., Neuhold, E. (eds.) FDSE 2016. LNCS, vol. 10018, pp. 305–316. Springer, Cham (2016). https://doi.org/10.1007/978-3-319-48057-2_21

One-Class Collective Anomaly Detection Based on LSTM-RNNs

Nga Nguyen Thi[1]([✉]), Van Loi Cao[2], and Nhien-An Le-Khac[2]

[1] Institute of Electronic, Institute of Military Science and Technology,
Hanoi, Vietnam
ngadtvt@gmail.com
[2] University College Dublin, Dublin, Ireland
loi.cao@ucdconnect.ie, an.lekhac@ucd.ie

Abstract. Intrusion detection for computer network systems has been becoming one of the most critical tasks for network administrators today. It has an important role for organizations, governments and our society due to the valuable resources hosted on computer networks. Traditional misuse detection strategies are unable to detect new and unknown intrusion types. In contrast, anomaly detection in network security aims to distinguish between illegal or malicious events and normal behavior of network systems. Anomaly detection can be considered as a classification problem where it builds models of normal network behavior, of which it uses to detect new patterns that significantly deviate from the model. Most of the current approaches on anomaly detection is based on the learning of normal behavior and anomalous actions. They do not include memory that is they do not take into account previous events classify new ones. In this paper, we propose a one-class collective anomaly detection model based on neural network learning. Normally a Long Short-Term Memory Recurrent Neural Network (LSTM RNN) is trained only on normal data, and it is capable of predicting several time-steps ahead of an input. In our approach, a LSTM RNN is trained on normal time series data before performing a prediction for each time-step. Instead of considering each time-step separately, the observation of prediction errors from a certain number of time-steps is now proposed as a new idea for detecting collective anomalies. The prediction errors of a certain number of the latest time-steps above a threshold will indicate a collective anomaly. The model is evaluated on a time series version of the KDD 1999 dataset. The experiments demonstrate that the proposed model is capable to detect collective anomaly efficiently.

Keywords: Long Short-Term Memory · Recurrent Neural Network
Collective anomaly detection

1 Introduction

Network anomaly detection refers to the problem of detecting illegal or malicious activities or events from normal connections or expected behavior of network

© Springer-Verlag GmbH Germany 2017
A. Hameurlain et al. (Eds.): TLDKS XXXVI, LNCS 10720, pp. 73–85, 2017.
https://doi.org/10.1007/978-3-662-56266-6_4

systems [3,5]. It has become one of the most popular subjects in the network security domain due to the fact that many organizations and governments are now seeking good solutions to protect valuable resources on computer networks from unauthorized and illegal accesses, network attacks or malware. Over the last three decades, machine learning techniques are known as a common approach for developing network anomaly detection models [2,3]. Network anomaly detection is usually posed as a type of classification problem: given a dataset representing normal and anomalous examples, the goal is to build a learning classifier which is capable of signaling when a new anomalous data sample is encountered [5].

Most of the existing approaches consider an anomaly as a discrete single data point: cases when they occur "individually" and "separately" [6,7,16]. In such approaches, anomaly detection models do not have the ability to represent the information from previous data points or events for evaluating a current point. In network security domain however, some kinds of attacks (e.g. *Denial of Service - DoS*) usually occur for a long period of time (several minutes) [10], and are often represented by a sequence of single data points. Thus these attacks will be indicated only if a sequence of single data points are considered as attacks. In this context, network data can be considered as a time series data that are sequences of events obtained over repeated measurements of time. Many approaches ranging from statistical techniques to machine learning techniques are employed for analyzing time series data, and their efficiency has been proven in time series forecasting problems. These approaches are based on the information of previous events to forecast the incoming step. In order to detect these kinds of attack mentioned above, anomaly detection models should be capable of memorizing the information from a number of previous events, and representing the relationship between them and with the current event. The model should not only have the ability to estimate prediction errors *anomalous scores* for each individual time-step but also be able to observe sequences of time-steps that are potential to be collective anomaly. To avoid important mistakes, one must always consider every outcome: in this sense a highly anomalous value may still be linked to a perfectly normal condition, and conversely. In this work, we aim to build an anomaly detection model for this kind of attacks (known as *collective anomaly detection* in [5]).

Collective anomaly is the term to refer to a collection of related anomalous data instances with respect to the whole dataset [5]. A single data point in a collective anomaly may not be considered as anomalies by itself, but the occurrence of a sequence of single points together may indicate a collective anomaly. Hidden Markov model, Probabilistic Suffix Trees, etc. are popular techniques for collective anomaly detection [5]. Recently, Long Short-Term Memory Recurrent Neural Network [8] has been recognized as a powerful technique to represent the relationship between a current event and previous events, and handles time series problems [12,14]. However, these approaches are proposed only for predicting time series anomalies at individual level (predicting prediction error *anomalous score* for each time-step), not at the collective level (observing prediction errors for a sequence of time-steps). Moreover, both normal and anomalous data are employed for training stage: either for training process (constructing classified models) or validation process (estimating model parameters). Thus, such the models are limited to detect new kinds of network attack. Collecting and labeling anomalous data are also expensive and time-consuming tasks. Therefore, we will propose a collective anomaly detection model by using the predictive power of LSTM RNN. This is to continue developing our idea proposed in [4]. The ability to detect collective anomaly of the proposed model will be demonstrated on *DoS* attack group in the KDD Cup 1999 dataset.

The rest of the paper is organized as follows. We briefly review some work related to anomaly detection and LSTM RNN in Sect. 2. In Sect. 3, we give a short introduction to LSTM RNN. This is followed by a section proposing the collective anomaly detection model using LSTM RNN. Experiments, Results and Discussion are presented in Sects. 5 and 6 respectively. The paper concludes with highlights and future directions in Sect. 7.

2   Related Work

When considering a time series dataset, point anomalies are often directly linked to the value of the considered sample. However, attempting real time collective anomaly detection implies always being aware of previous samples, and more precisely their behavior. This means that every time-step should include an evaluation of the current value combined with the evaluation of preceding information. In this section, we briefly describe previous work applying LSTM RNN for time series and collective anomaly detection problems [12,14,15,17].

Olsson et al. [15] proposed an unsupervised approach for detecting collective anomalies. In order to detect a group of the anomalous examples, the *anomalous score* of the group of data points was probabilistically aggregated from the contribution of each individual example. Obtaining the collective anomalous score was carried out in an unsupervised manner, thus it is suitable for both unsupervised and supervised approaches to scoring individual anomalies. The model was evaluated on an artificial dataset and two industrial datasets, detecting anomalies in moving cranes and anomalies in fuel consumption.

In [12], Malhotra et al. applied a LSTM network for addressing anomaly detection problem in time series fashion. A stacked LSTM network trained on only normal data was used to predict values of a number of time-steps (L steps) ahead. Thus, the prediction model produced L prediction values in the period of L time-steps for a single data point. It then resulted in a prediction error vector with L elements for each data point. The prediction error of a single point was then computed by modeling its prediction error vector to fit a multivariate Gaussian distribution, which was used to assess the likelihood of anomaly behavior. Their model was demonstrated to perform well on four datasets.

Marchi et al. [13,14] presented a novel approach by combining non-linear predictive denoising autoencoders (DA) with LSTM for identifying abnormal acoustic signals. Firstly, LSTM Recurrent DA was employed to predict auditory spectral features of the next short-term frame from its previous frames. The network trained on normal acoustic recorders tends to behave well on normal data, and yields small reconstruction errors whereas the reconstruction errors from abnormal acoustic signals are high. The reconstruction errors of the autoencoder was used as an "anomaly score", and a reconstruction error above a predetermined threshold indicates a novel acoustic event. The model was trained on a public dataset containing in-home sound events, and evaluated on a dataset including new anomaly events. The results demonstrated that their model performed significantly better than existing methods. The idea is also used in a practical acoustic example [13,14], where LSTM RNNs are used to predict short-term frames.

In [17] Ralf et al. employed LSTM-RNN for intrusion detection problem in supervised manner. The processed version of the KDD Cup 1999 dataset, which is represented in time-series, were fed to the LSTM-RNN. The network has five outputs representing the four groups of attacks and normal connections in the data. Both labeled normal connections and labeled attacks were used for training the model and estimating the best LSTM-RNN architecture and its parameters. The selected model was then evaluated on 10% of the corrected dataset under measurements of confusion matrix and accuracy. The results shown that their model performed well in terms of accuracy, especially it achieved very high performance on two groups of attacks, Probe and DoS.

To the best of our knowledge, non of previous work using LSTM RNN addresses the problem of collective anomaly detection. Thus, we aim to develop collective anomaly detection using LSTM RNN. Our solution consists of two stages: (1) LSTM RNN will be employed to represent the relationship between previous time-steps and current one in order to estimate *anomalous score* (known as prediction error) for each time-step. This stage is considered as developing a time series anomaly detection, and similar to the previous work [12]; (2) A method will be proposed for observing sequences of single data points based on their anomalous scores to detect collective anomaly. The second stage makes our work original and different from previous work that applied LSRM RNN for time series anomaly detection. This will prove very efficient in our example: First, we will train an LSTM RNN on only normal data in order to learn the normal behavior. The trained model will be validated on normal validation set in order to estimate the model parameters. The result classifier will then employ to rate the *anomalous score* for data at each time-step. The *anomalous score* of a sequence of time steps will be aggregated from the contribution of each individual one. By imposing a predetermine threshold a sequence of single time-steps will indicate as a collective anomaly if its anomalous score is higher than the threshold. More details on our approach can be found in Sect. 4.

3   Preliminaries

In this section we briefly describe the structure of Long Short Term Memory nodes, and the architecture of a LSTM RNN using LSTM hidden layer. The LSTM was proposed by Hochreiter et al. [8] in 1997, and has already proven to be a powerful technique for addressing the problem of time series prediction.

The difference initiated by LSTM regarding other types of RNN resides in its "smart" nodes presented in Hidden layer block in Fig. 1. Each of these cells contains three gates, input gate, forget gate and output gate, which decide how to react to an input. Depending on the strength of the information each node receives, it will decide to block it or pass it on. The information is also filtered with the set of weights associated with the cells when it is transferred through these cells.

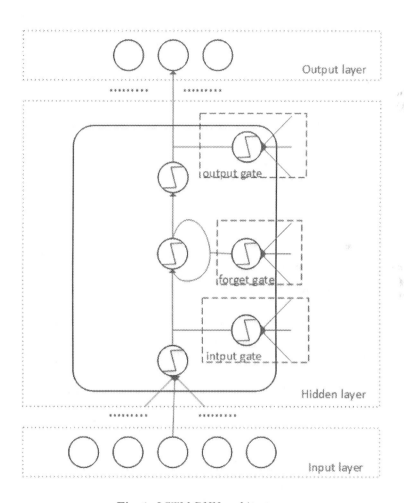

Fig. 1. LSTM RNN architecture

The LSTM node structure enables a phenomenon called backpropagation through time. By calculating for each hidden layer the partial derivatives of the output, weight and input values, the system can move backwards to trace the evolving error between real output and predicted output. Afterwards, the network uses the derivative of this evolution to adapt its weights and decrease prediction error. This learning method is named Gradient descent.

A simple LSTM RNN as in Fig. 1 consists three layers: Input layer, LSTM hidden layer and output layer. The input and output layers are the same as those in multi-layered perceptrons (MLPs). The input nodes are the input data, and the output nodes can be sigmoid, tanh or other transform functions. The LSTM hidden layer is formed from a number of the "smart" nodes that are fully connected to the input and output nodes. Two common techniques, Gradient descent and Back-propagation can be used for optimizing its loss function and updating its parameters.

As mentioned before, Long Short-Term Memory has the power to incorporate a behaviour into a network by training it with normal data. The system becomes representative of the variations of the data. In other words, a prediction is made focusing on two features: the value of a sample and its position at a specific time. This means that two same input value at different times probably results in two different outputs. It is because a LSTM RNN is stateful, i.e. has a "memory", which changes in response to inputs.

4   Proposed Approach

As mentioned in the Related work section, one of the most recent research using LSTM RNNs for building anomaly detection model in time series data is from [12]. The model was demonstrated to be efficient for detect anomalies of time series data. In their model, the prediction errors of a data point is computed by fitting its prediction errors vector to a multivariate Gaussian distribution which is then used to assess the likelihood of anomalous behavior. Hence, it is only suitable for detecting abnormal events that happen instantly i.e. in a very short time such as in the electrocardiogram or power demand applications since the model do not have the ability of estimating likelihood anomaly for a long period of time (a sequence of data points). Consequently, the model is not suitable for the collective anomaly detection in the context of network security where some kinds of network attack last for a significant period time.

Therefore, in this paper we propose a new approach of using LSTM RNNs for cyber-security attacks at collective level. We will use a simple LSTM RNN architecture, in contrast to a stacked LSTM in [12]. This does not change the core principle of the method: when given sufficient training, a LSTM RNN adapts its weights, which become characteristic of the training data. In the network, each output nodes represents each time-step that the model predicts. For example, if the model is fed a data point at time-step t-th for predicting values at next three steps ahead, the first, second and third output nodes will represent the values at time-steps $(t + 1)$-th, $(t + 2)$-th and $(t + 3)$-th respectively. Instead of fitting a mixture Gaussian model, we simple compute the prediction error of a individual data point by mean over the prediction errors at three time-steps. We show the LSTM RNNs ability to learn the behavior of a training set, and in this stage it acts like a time series anomaly detection model. Following this stage, we proposed terms (defined below) to monitor the prediction errors of a certain number of successive data points and locate collective anomaly in data. The second stage allows our model to detect collective anomaly, and make model original and different from previous ones. Thus, the performance of our model is not compared to that of previous models using LSTM RNN on time series anomaly detection, e.g. [12].

Terms for measuring prediction errors of data point, and monitoring anomalous behavior in a period of time-steps are defined as below:

- **Relative Error (RE):** the relative prediction error between a real value x and its prediction value $\hat{x}$ from LSTM RNN at every time-step is defined as in Eq. 1. Note that a single data can not be considered as a collective anomaly. However, the larger value of RE a data point has, the higher probability the single data point belongs to a collective anomaly.

$$RE\,(x, \hat{x}) = |x - \hat{x}| \tag{1}$$

- **Prediction Error Threshold (PET):** It is employed to determiner an individual query point (a time-step) can be classified as a normal time-step or considered as an element in a potential collective anomaly. Its prediction error RE above threshold PET may indicate a element in a collective anomaly.
- **Collective Range (CR):** a minimum number of anomalies appearing successively in a network flow are considered as a collective anomaly. Both PET and CR are estimated based on the best classification performance of the model on normal validation set.

5 Experiments

5.1 Datasets

In order to demonstrate the efficient performance of the proposed model, we choose a dataset related to the network security domain, the KDD 1999 dataset [1,9], for our experiments. The dataset in *tcpdump* format was collected from a simulated military-like environment over a period of 5 weeks (from 1st March 1999 to 9th April 1999). The dataset is composed of a two-weeks for training, weeks 1 and week 3 (free attack), one week for validation, week 2 (labeled attacks), and other two weeks for testing, weeks 4 and 5 (both normal and anomalous data).

There are four main groups of attacks in the dataset, but we restrict our experiments on a specific attack, *Neptune*, in the *Denial-of-Service* group. The dataset is also converted into a time series version before feeding into these models. More details about how to obtain a time series version from the original *tcpdump* dataset, and how to choose training, validation and testing sets are presented in the following paragraphs.

The first crucial step is to build a conveniently usable time series dataset out of the tcpdump data, and to select the interested features. We use terminal commands and a python program to convert the original tcpdump data in the KDD 1999 dataset into a time dependent function. This method is a development of the proposed transformation in [11] that acts directly on the tcpdump to obtain real time statistics of the data. Our scheme follows this step by step transition as described below:

$$\text{tcpdump} \Rightarrow \text{pcap} \Rightarrow \text{csv}$$

Each day of records can be time-filtered and input into a new *.pcap* file. This also has the advantage of giving a first approach on visualizing the data by using Wireshark functionalities. Once this is done, the *tshark* command is adapted to select and transfer the relevant information from the records into a *csv* format. We may note that doing this is a first step towards faster computation and better system efficiency, since all irrelevant pcap columns can be ignored. Although the method for converting data do not suit for the model performing real-time, it is sufficient enough for evaluating the proposed model in detecting collective anomaly as the main objective of this work. However, if the attack data is recorded in real-time under time series format, then our method can be applied in real-time detection.

In our experiments, we use 6-days normal traffic in the first and third weeks for training, n_{train}, and one-day normal traffic (Thursday) in week 3 for validation, n_{valid}. Testing sets include 1-day normal traffic from week 3 (Friday), n_{test}, and 1-day data containing attacks in week 2 (Wednesday), a_{test}. The protocol will be the following: training the network with n_{train}, using n_{valid} for choosing *Prediction Error Threshold (PET)* and *Collective Range (CR)*, and evaluating the proposed models on n_{test} and a_{test}.

Table 1. Parameter settings

LSTM RNN parameters	
Input size	1
Hidden LSTM layer	10
Output sigmoid layer	1, 2 or 3
Learning rate	10^{-4}
Number of epochs	100
Momentum	0.5
Batch size	1
Collective thresholds	
Prediction Error Threshold (PET)	0.3
Collective Range (CR)	4

5.2   Experimental Settings

Our experiments are aim to demonstrate the ability of detecting collective anomalies of the proposed model. However, there is no anomalous instances available for training and validation stages, it is much harder than binary classification problem in estimating hyper-parameters, optimizing network architectures and setting thresholds. Thus, we will briefly discuss this issues in next paragraph, and design two experiments, a preliminary experiment for choosing these thresholds and a main experiment for evaluating the proposed model.

We investigate three network architectures. The difference amongst these architectures is only the size of output layer, with one, two and three outputs for predicting 1-step, 2-step and 3-step ahead respectively. This means that the three-output network can predict three values for three steps ahead with a current input. The number of hidden nodes and the learning rate can strongly influence the performance of a LSTM RNN. Each synapse of a network is weighted differently, and can be considered as a unique interpretation of the input data. Each node of the hidden layer is storage space for these interpretations. Theoretically, the higher number of hidden nodes, the more information the network can contain. This also means more computation, and may lead to over-fitting. Therefore, it should be trade off between the detection rate and the computational expenses on constructing models and querying new data. In this paper, we choose the size of LSTM hidden layer equal to 10. The learning rate is another factor directly linked to the speed at which a LSTM RNN can improve its predictions. For a time step t during training process, the synapse weights of the neural network are updated. The learning rate defines how much we wish a weight to be modified at each instant. We choose a common used value for learning rate, 10^{-4}.

The preliminary experiment aim to tune *Prediction Error Threshold (PET)* and *Collective Range (CR)* by using only normal validation set, n_{valid}. This means that these parameters are choose so as to the model can correctly classify most of instances in n_{valid} (say 95%, 97% or 100%) belonging to normal class. In this paper, we set these threshold so as to keep 100% of examples in n_{valid} as normal data. The choice of CR depends on how long a single data point represents for. In our data, each single point represents a period of 10 min, so CR. Thus we choose CR equal to 4 which is equivalent to a period of 40 min. One CR chosen, we will compute detection rate of the model with 20 different values of PET ranging from 0.05 to 1.0. on n_{valid}. The smallest value of PET that enables the model to correctly classify 100% examples in n_{valid} is chosen. More details about network architectures, parameters settings are presented in Table 1.

The main experiment is to shown the ability of LSTM-RNNs in detecting a disproportionate durable change in a time series anomaly. Once the preliminary experiment is complete, these trained models with collective thresholds, PET and CR is employed to detect anomalous region in data, n_{valid}, n_{test} and a_{test}. This experimental results include the prediction error of each single data point, specific anomalous regions on normal validation set and test sets from the three models. The training error is plotted in Fig. 2. Figures 3, 4 and 5 illustrate the prediction errors on validation set (v_n), testing sets (n_{test} and a_{test}). Specific regions predicted as collective anomaly and the proportion of these anomalous regions are presented in Table 2.

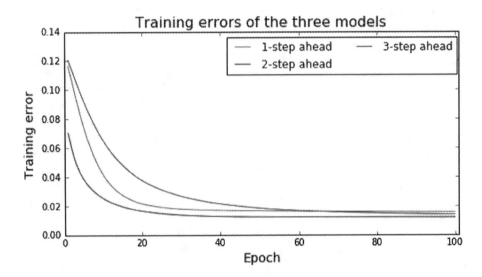

Fig. 2. The training errors of the proposed model

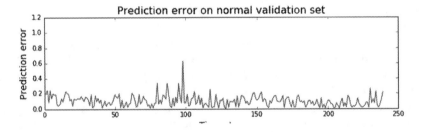

Fig. 3. The prediction error from 3-step ahead model on validation set.

6   Results and Discussion

The Table 2 shows the collective anomaly prediction of the proposed model on three datasets, n_{valid}, n_{test} and a_{test}. The collective anomaly prediction includes specific regions in data and the percentage of data instances (time-steps) within these regions. There is no anomalous region found in normal validation set, n_{valid} because the thresholds, PET and CR have been tuned to classify 100% of n_{valid} belonging to normal class. Only one regions in normal test, n_{test} is found by the third classifier (3-step ahead), which is False Negative. In anomaly test, many regions are detected as collective anomalies by both of three classifiers. It seems

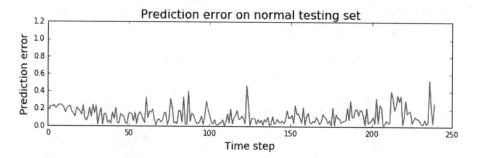

Fig. 4. The prediction error from 3-step ahead model on normal test.

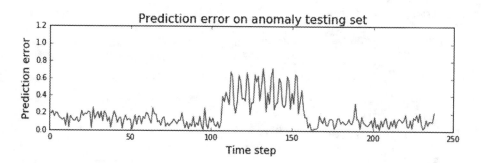

Fig. 5. The prediction error from 3-step ahead model on anomaly test.

Table 2. The prediction collective anomalies in validation and test sets

Dataset	1-step ahead		2-step ahead		3-step ahead	
	Anomaly region	Anomaly ratio	Anomaly region	Anomaly ratio	Anomaly region	Anomaly ratio
n_{valid}	-	0.0%	-	0.0%	-	0.0%
n_{test}	-	0.0%	-	0.0%	215–219	1.67%
a_{test}	107–111	8.79%	107–111	12.13%	107–111	14.23%
	116–124		116–124		116–124	
	125–134		125–134		125–134	
	-		135–139		135–139	
	-		140–144		140–145	
	-		-		150–154	

to be that the more steps ahead the model can predict, the more regions can be found. They are three regions (8.79%), five regions (12.13%) and six regions (14.23%) found by the first, second and third classifiers respectively.

The Fig. 2 illustrates the training errors from three classifiers with 1-step, 2-step and 3-step prediction ahead. These errors tend to converge after 100 epochs, and the error curve of the 3-step classifier levels off earlier than two others. The Figs. 3, 4 and 5 also present the prediction errors of the third classifiers on n_{valid}, n_{test} and a_{test}. In Figs. 3 and 4, the prediction errors are just fluctuated around 0.2, and few individual errors have large values. Thus, these datasets are not considered as collective anomaly. However, the error patterns in Fig. 5 are quite different. The errors within time steps 110 to 155 is quite high, much higher than the rest of error regions. This regions are detected as collective anomalies as presented in Table 2.

In training stage, the more prediction time-steps a model has, the less training error the model produces (see Fig. 2). This suggests that the models with more prediction time-steps tend to learn the normal behaviors of network traffic more efficiently than ones with less prediction time-steps. However, the number of prediction time-steps can enrich the model's ability to detect anomaly regions in the anomaly testing set a_{test}. This may imply that the three-steps model is more robust to learn the normal behaviors and identify collective anomalies than the two others.

7   Conclusion and Further Work

In this paper, we have proposed a model for collective anomaly detection based on Long Short-Term Memory Recurrent Neural Network. We have motivated this method through investigating LSTM RNN in the problem of time series, and adapted it to detect collective anomalies by proposing the measurements in Sect. 4. We investigated the hyper-parameters, the suitable number of inputs and some thresholds by using the validation set.

The proposed model is evaluated by using the time series version of the KDD 1999 dataset. The results suggest that proposed model is efficiently capable of detecting collective anomalies in the dataset. However, they must be used with caution. The training data fed into a network must be organized in a coherent manner to guarantee the stability of the system. In future work, we will focus on how to improve the classification accuracy of the model. We also observed that implementing variations in a LSTM RNNs number of inputs might trigger different output reactions.

References

1. DARPA intrusion detection evaluation. (n.d.). http://www.ll.mit.edu/ideval/data/1999data.html. Accessed 30 June 2016
2. Ahmed, M., Mahmood, A.N., Hu, J.: A survey of network anomaly detection techniques. J. Netw. Comput. Appl. **60**, 19–31 (2016)

3. Bhattacharyya, D.K., Kalita, J.K.: Network Anomaly Detection: A Machine Learning Perspective. CRC Press, Boca Raton (2013)
4. Bontemps, L., Cao, V.L., McDermott, J., Le-Khac, N.-A.: Collective anomaly detection based on long short-term memory recurrent neural networks. In: Dang, T.K., Wagner, R., Küng, J., Thoai, N., Takizawa, M., Neuhold, E. (eds.) FDSE 2016. LNCS, vol. 10018, pp. 141–152. Springer, Cham (2016). https://doi.org/10.1007/978-3-319-48057-2_9
5. Chandola, V., Banerjee, A., Kumar, V.: Anomaly detection: a survey. ACM Comput. Surv. (CSUR) **41**(3), 15 (2009)
6. Chmielewski, A., Wierzchon, S.T.: V-detector algorithm with tree-based structures. In: Proceedings of the International Multiconference on Computer Science and Information Technology, Wisła, Poland, pp. 9–14. Citeseer (2006)
7. Hawkins, S., He, H., Williams, G., Baxter, R.: Outlier detection using replicator neural networks. In: Kambayashi, Y., Winiwarter, W., Arikawa, M. (eds.) DaWaK 2002. LNCS, vol. 2454, pp. 170–180. Springer, Heidelberg (2002). https://doi.org/10.1007/3-540-46145-0_17
8. Hochreiter, S., Schmidhuber, J.: Long short-term memory. Neural Comput. **9**(8), 1735–1780 (1997)
9. KDD Cup Dataset (1999). http://kdd.ics.uci.edu/databases/kddcup99/kddcup99.html
10. Lee, W., Stolfo, S.J.: A framework for constructing features and models for intrusion detection systems. ACM Trans. Inf. Syst. Secur. (TiSSEC) **3**(4), 227–261 (2000)
11. Lu, W., Ghorbani, A.A.: Network anomaly detection based on wavelet analysis. EURASIP J. Adv. Sig. Process. **2009**, 4 (2009)
12. Malhotra, P., Vig, L., Shroff, G., Agarwal, P.: Long short term memory networks for anomaly detection in time series. In: Proceedings, p. 89. Presses universitaires de Louvain (2015)
13. Marchi, E., Vesperini, F., Eyben, F., Squartini, S., Schuller, B.: A novel approach for automatic acoustic novelty detection using a denoising autoencoder with bidirectional LSTM neural networks. In: 2015 IEEE International Conference on Acoustics, Speech and Signal Processing (ICASSP), pp. 1996–2000. IEEE (2015)
14. Marchi, E., Vesperini, F., Weninger, F., Eyben, F., Squartini, S., Schuller, B.: Non-linear prediction with LSTM recurrent neural networks for acoustic novelty detection. In: 2015 International Joint Conference on Neural Networks (IJCNN), pp. 1–7. IEEE (2015)
15. Olsson, T., Holst, A.: A probabilistic approach to aggregating anomalies for unsupervised anomaly detection with industrial applications. In: FLAIRS Conference, pp. 434–439 (2015)
16. Salama, M.A., Eid, H.F., Ramadan, R.A., Darwish, A., Hassanien, A.E.: Hybrid intelligent intrusion detection scheme. In: Gaspar-Cunha, A., Takahashi, R., Schaefer, G., Costa, L. (eds.) Soft Computing in Industrial Applications. AISC, vol. 96, pp. 293–303. Springer, Heidelberg (2011). https://doi.org/10.1007/978-3-642-20505-7_26
17. Staudemeyer, R.C., Omlin, C.W.: Evaluating performance of long short-term memory recurrent neural networks on intrusion detection data. In: Proceedings of the South African Institute for Computer Scientists and Information Technologists Conference, pp. 218–224. ACM (2013)

Multihop Wireless Access Networks for Flood Mitigation Crowd-Sourcing Systems

Quang Tran Minh[1($\boxtimes$)] and Michel Toulouse[2]

[1] Ho chi minh City University of Technology, VNU-HCM,
268 Ly Thuong Kiet, Ho Chi Minh City, Vietnam
`quangtran@hcmut.edu.vn`
[2] Faculty of Engineering, Vietnamese German University,
Ho Chi Minh City, Vietnam
`michel.toulouse@vgu.edu.vn`

Abstract. Natural disasters can be mitigated or even anticipated if we have appropriate means, in terms of communications and data sharing models, to collect relevant data in advance or during disaster occurrences, which can be used for supporting disaster prevention and recovery processes. This work proposes a framework that encourages people to collect and share data about disaster, especially flood in Ho Chi Minh City, via on-site established multihop wireless access networks configured by the sharing of internet connectivity in users' mobile devices. For connectivity sharing, on-the-fly establishment of multihop wireless access network (OEMAN) scheme is thoroughly analyzed and improved to resolve its inherent issue on traffic load imbalance due to its tree-based structure. More specifically, we propose a linear program for overload-aware routing optimization considering wireless interference. Evaluations implemented in Matlab show that the overload-aware routing improves load balancing among available virtual access points in OEMAN. By avoiding nodes with heavy load in the network, our solution improves network throughput compared to overload-unaware routing protocols.

Keywords: Overload-aware routing · Network load balancing
Linear programming · Crowd-sourcing · Incentive schemes

1 Introduction

Natural disasters such as flood, fire, earthquake, hurricane, cyclone, volcanic eruption, etc., often cause devastating damages to goods, buildings, equipments, infrastructures, and even loss of human lives [1]. These disasters can be mitigated or even anticipated if an appropriate information communication technology (ICT) framework was made available. In flood situations, such system could collect and analyze relevant data to provide useful information such as flood forecast, drawing an online map that updates the effects of a flood to avoid unexpected damage and finding afflicted locations for quick emergency relief. Unfortunately, such an intelligent ICT system is not yet available.

© Springer-Verlag GmbH Germany 2017
A. Hameurlain et al. (Eds.): TLDKS XXXVI, LNCS 10720, pp. 86–108, 2017.
https://doi.org/10.1007/978-3-662-56266-6_5

This work aims at proposing a framework to encourage people to collect and share flood data. Data related to a flood such as the location, flooding level, time, picture of the effected area, and so on can be collected by mobile devices such as mobile phones and tablets which are always available on-site. These data can then be transmitted to a data center for analysis, providing useful information to relevant users such as citizen living in or traveling to the affected areas, as well as to the disaster emergency units for coordination of the mitigation and recovery efforts.

Communications infrastructures are often seriously disable when major natural disasters occur, as observed during the Kumamoto earthquakes of Japan in April 2016 [2], the Northeast India earthquake in January 2016 [3] or the Koppu Typhoon in the Philippines in October 2015 [4]. Services like Internet connectivity could be lost in disaster areas while in the same time there might be mobile devices still functioning, capable of helping rescuers to localize and access victims trapped in particular disaster areas such as flood zones. In such situation, impacted users may still be able to share data locally and if someones in the area still has Internet connectivity then local data related to flood and its effect can be made available to emergency services.

In [5], the authors proposed OEMAN, a new system for the establishment of multihop wireless access networks in disaster affected areas. OEMAN takes advantage of existing mobile devices in a disaster area to extend Internet connectivity by creating virtual access points (VAPs) on functioning mobile devices. After the establishment of the virtual network, victims can connect to actual access points (APs) which are still-alive in a disaster area via intermediate VAPs of OEMAN, to then access the Internet and inform rescuers about their status. This system can be utilized for flood mitigation as well, thereby users can share the flood related information locally via multihop communications or globally via the Internet connectivity shared by nearby devices.

A system like OEMAN should not require the intervention of the victims in its set up or activation phases. In [6], the authors added a new set of functions in their system, namely wireless multihop communication abstraction (WMCA), so that the system can be more autonomous. WMCA allows any OEMAN node to connect to another network as a station (STA) for its Internet access and to share its connectivity as VAP for the nearest OEMAN nodes. A wireless virtualization [7] is deployed to switch the WiFi interface (WIF) between STA mode and VAP mode. Figure 1 shows a simple WMCA in a disaster region. Figure 1(a) shows that mobile nodes (MNs) 2 and 3 fail to keep its connect with an existing AP as the AP is damaged by a disaster. Meanwhile, Fig. 1(b) reveals that if each MN can provide a connectivity to the nearby nodes leveraging the WMCA mechanism then the Internet connectivity can be extended to nodes in the disaster affected area allowing victims to access to the Internet and share their safety information.

The system proposed in [5,6] provides a suitable environment for users to share data and support each other in mitigating the negative effects of a natural disaster. However, this scheme still has several limitations that need to be

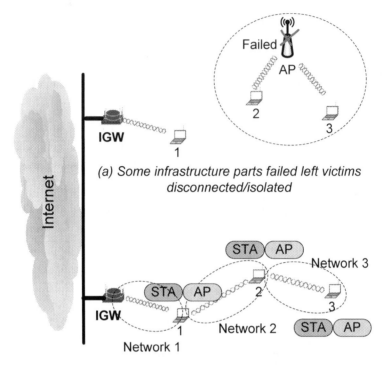

(a) Some infrastructure parts failed left victims
disconnected/isolated

(b) Extending Internet connectivity from still alive nodes

Fig. 1. Wireless multihop communication abstraction (WMCA).

addressed, such as handling mobility and node failure, load balancing, multiple
surviving APs and power consumption issues [5]. According to our analysis, load
balancing is one of the most important issue which should be resolved for maintaining the network stability. OEMAN [5,6] greedily generates tree-based access
networks which easily become imbalanced with bottlenecks at nodes which are
close to the root (i.e., AP serving as an Internet gateway - IGW). This imbalance
may lead to network failures. Meanwhile, other APs may still survive providing
chances for better Internet connectivity provision, opportunities which have not
yet been addressed in OEMAN.

Two approaches have been considered to enhance the capability of OEMAN
over multiple IGWs: (i) creating multiple separated networks, each rooted from
a particular IGW; (ii) creating overlapping networks where nodes could be part
of multiple networks at the same time. The former still inherits the bottleneck
issue in the original OEMAN. The later is much less prone to bottleneck issues as
each node can dynamically identify an adequate IGW as well as the appropriate
route to such an IGW for data forwarding, considering network conditions such
as the load of nodes on each route, wireless interference, and so on.

The main contributions in this paper are as follows. We propose a framework for flood mitigation which collects data from the crowd in order to provide updated information for quick emergency responses and mitigation. A simple but yet effective crowd-sourcing model is proposed leveraging communications means provided by OEMAN for sharing flood related data in a severe environment, where main parts of communications infrastructures might already be destroyed. As mentioned already, load balancing is one essential issue in OEMAN which has not been thoroughly resolved. A solution is proposed to reduce end to end (E2E) communications delays and to increase the system throughput by balancing the load among VAPs in the OEMAN system. More specifically, to distribute traffic among VAPs and achieve a better load balancing, we propose a linear programming model which maximizes the system throughput by allocating traffic through multipath routing while avoiding communications interference, a common issue in wireless communications. We have implemented our linear programming model in Matlab. Simulation results show benefits in terms of throughput and E2E delays of the overload-aware routing compared with current overload unaware-routing in OEMAN.

The remainder of this paper is organized as follows. Section 2 introduces works closely related to our work. Section 3 provides the overall architecture of our flood mitigation system which can be applied for example to an area like Ho Chi Minh City. Section 4 introduces our proposed scheme for sharing data and communications means when disaster, especially flood, occurs. Section 5 describes our optimization model for overload-aware routing applied to the OEMAN scheme for disaster recovery access networks. Section 6 presents and analyzes the results, in terms of routing optimization, using simulations in Matlab. Finally, we concludes this paper and discus future work directions in Sect. 7.

2   Related Work

Climate changes associated to global warming are responsible for an increase in the frequency and severity of floods occurring all over the world. These floods often have devastating effects, particularly in tropical countries, including Vietnam, in regions like Ho Chi Minh City, the Mekong Delta and the central provinces such as Binh Dinh and Quang Nam. Various researches have been conducted on flood mitigation and associated risk management [8,9]. To mitigate the impacts of a flood, authorities need proper monitoring systems that provide frequent updates about the flood status. Display of such information on an online map can help provide suitable instructions for evacuation and coordinate risk mitigation and quick responses by rescuer teams [10]. Accurate and up-to-date information about a flood can be obtained from different sources such as sensor systems, satellites, mobile devices from crowds and so on [11].

Conventional flood risk management systems such as the information portal for flood mitigation at the Flood Mitigation Center - HCMC [8] rely mainly on observation stations with rain gauges, tide or river level measurements deployed at fix-locations. Unfortunately such systems do not report detailed information about flood impacts such as affected locations/streets, granular flooding levels, affecting durations, real damage/loss on goods and lives, etc. As an effort to overcome these drawbacks, Do et al. [12] have proposed a system that automatically collects the related information to provide early flood warnings. This system uses mobile devices to sense the rain volume and river level, reporting these data to the center using GPRS (General Packet Radio Service) technology. This system has some limitations in terms of implementation cost as well as drawbacks in terms of communications technology since GPRS is not suitable for multimedia data communications such as pictures or videos which are much needed for reporting flood impacts.

Most of the existing work for disaster recovery, including flood mitigation systems, focus on data acquisition, communications and data analysis issues. For data acquisition, wireless sensor networks have been used for automatically collecting data in a vast area [13–15]. For communications between the data acquisition systems and the data analysis centers in severe flood disaster environments, wireless and mobile communications have been leveraged [5,6]. Data analysis centers employ data analytic techniques such machine learning, data mining, artificial neural network and so on for early flood detection, and impact analysis to suggest suitable mitigation methods [16].

With advances in wireless and mobile technologies such as 4G/5G, WiMax, LTE, etc., as well as smart mobile devices such as smartphones, tablets, etc., data collection based on contributions from communities is increasingly becoming a feasible and relevant approach [17,18]. A difficulty however with this approach is the design of an appropriate incentive model which encourages crowds to collect and share data via their mobile devices. Our research proposes an effective incentive model based on game theory which encourages individuals in crowds to collect and share data [19].

Another core difficulty in building efficient disaster recovery systems including the flood risk management and mitigation systems comes from communications aspects. Whether it is residual capacities from an existing but degraded network or network capabilities leverage from mobile devices such as in OEMAN, the still available communication resources should be used fully to provide the communication services always in great demand in such circumstances. This problem of maximizing the usage of communication resources can be seen as a resource-allocation optimization problem. Solutions to such problem seek to optimize some specific parameters of the available network capabilities such as minimizing packets lost or delay, or maximizing network throughput. In this work we address one of the problems in OEMAN, network load imbalance, through this angle. Network imbalances reduce the amount of traffic that can be transferred successfully. In this study we propose a linear programming model to maximize network throughput defined as the amount of traffic transferred successfully over a communications channel in a given period of time (e.g., in a second). Our linear model solves this optimization problem through multipath routing, thus distributing the traffic flow across a larger number of alive network nodes.

Several linear programming models to optimize throughput in wireless networks have been proposed already. Our model derives mostly from applications in network coding [20]. Network coding uses intermediate nodes to combine native packets into *coded packets* which are then broadcast to next hops. Network coding models create opportunities for coding by driving network flows towards same nodes, but this approach obviously creates bottlenecks [21] (similar to the bottleneck issue in OEMAN). This issue has been addressed in [22] by considering the best paths which not only have many coding opportunities but also achieve a load balancing among nodes in the network. We combine this approach with multipath routing [23] which consists of methods to route traffic from a source node to a destination node over several paths in a network. In wireless networks, multipath routing helps to improve network capacity utilization, load balancing, to reduce packet loss and delay. Multipath routing is also part of network coding [24] - where the traffic flows on different paths in order to find more coding chances.

Multipath routing raises the problem of signal interference, a major issue in multi-hop wireless networks. In general, one seeks to minimize interference for network functions like routing. An interference-aware multipath routing [25] has been proposed as a linear optimization program. With this model, we can find the paths which are away from each other for avoiding wireless interference.

3   Overall Architecture

The overall architecture of the proposed framework is depicted in Fig. 2 and described as follows. As shown on the left of Fig. 2, data related to a flood such as locations (e.g., latitude and longitude by Global Positioning System - GPS), flooding levels, time, pictures depicting the flood impact areas, and even individual comments (e.g., in text) etc., are collected from crowds using a crowdsourcing platform running on mobile devices such as mobile phones or tablets, which are always available on-site. These data are then transmitted to and analyzed at the data center (e.g., Data warehousing & analysis). The communications means between the crowds and the data center could be Wifi, 3G/4G, LTE, WiMax. In case of severe flood damaging the main parts of the network infrastructures, the OEMAN [5] scheme can be utilized. At the data center, flood data is well analyzed to provide useful information, from detailed data to multi-dimensional statistical data.

This system is useful to multiple stakeholders: citizen, management, users (people who shares data). For instance, the analyzed information can help citizen and management to early detect, estimate the effect of a flood thereby they can conduct tactical mitigation and recovery activities. In addition, urban management can utilize such information for strategic urban plannings for long term flooding avoidance plans. Beside that, users of the system could receive relevant compensations such as respect from community, money, or free-of-charge for added-value services from the proposed system (e.g., free of-charge for information about any flood in the whole city).

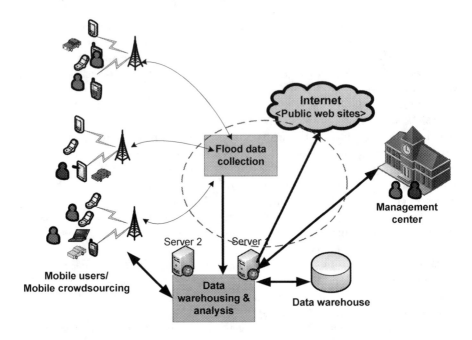

Fig. 2. The overall architecture of the proposed framework

In this work, we focus on the data collection and sharing part (shown in the left part of Fig. 2). As mentioned, flood may seriously degrade communications infrastructures, isolating people in the flood affected areas. In such scenario, the OEMAN system [5] can be used to extend the connectivity from still alive access points to otherwise isolated victims. However, there are still essential issues which need to be resolved before this system can be effectively deployed: (i) an incentive model which encourages users sharing flood related data and their mobile devices as communications means for Internet connectivity extension; (ii) OEMAN is vulnerable to load balancing issues as heavy load is accumulated at nodes near to the root nodes (i.e., the Internet gateway - IG). The issue (i) is briefly addressed in the next section while the issue (ii) will be thoroughly analyzed in the following sections.

4   Incentivizing to Share Data and Communication Means

In order to collect flood related data from the crowds, we need mobile users' willingness on sharing (a) the data itself (collected by users' mobile devices) and, (b) the communications means, i.e. WiFi network interface to extend a mobile device's connectivity to the nearby devices.

For (a), users must collect the flood related data including *locations, time, impact level, pictures representing the flood at the considering area,* as well as their *individual comments*. In order to improve willingness of users, we provide a crowd-sourcing platform to support users entering data in the most convenient way. Some data can be automatically collected by the platform such as *locations, time*. Other, such as the impact level is predefined by the platform as *serious, medium*, and *light* impacts, users only need to select one of the options in the drop box implemented in the platform. Users can also provide further details such as illustrating pictures (captured by the camera embedded on the devices) or comments expressing users' feeling about the flood. Although it does not require much effort from users to run such a system/platform, commonly users will not run this system for data collection if there is no clear benefit to them. Therefore, we need an incentive model to encourage users to collect and share the flood related data.

As for (b), OEMAN [5] provides a mechanism to share the connectivity of mobile devices easily without much involvement from users. In OEMAN, devices can detect the available Internet connectivity shared by the still alive nodes near by. However, sharing Internet connectivity will drain device's battery and expose users to security risk. Therefore, we also need an incentive model to encourage users to share connectivity on their devices.

In this work, we consider a *flood mitigation community network* launched by the Flood Management Center (ref. Fig. 2) using our proposed system (i.e., the crowd-sourcing platform combined with OEMAN). The network consists of a set of users, where each one owns a mobile device that can install our provided flood crowd-sourcing platform (to share flood related data and received flood related information such as information about the flood impact level at the considered location and evacuation instructions) and can function as a node in OEMAN (to share communications means). The network has two types of users: Subscribers and Aliens [19]. Subscribers are those willing to share data and their devices as mobile relays, whereas Aliens are those that only want to use the proposed system without sharing data or connectivity.

Similar as [19] and FON [26], we propose to incentive users to share data and connectivity via two different incentive schemes, corresponding to two types of memberships to subscribers, namely *Linus* and *Bill*. Linus is a member who does not receive any compensation when other users access his shared connectivity and data. Meanwhile, Linus can access flood related information from the system anytime needed and at any location, free of charge. A Bill member is a user that receives compensation when other users utilize his shared connectivity for Internet access. A Bill user also receive compensation related to the amount, the accuracy and timeliness of the data shared with the system for flood analysis. Meanwhile, a Bill member needs to pay for accessing flood related information from the system and for using the Internet connectivity shared by other users. If a user does not register to the proposed system, he can still access the system as an Alien (a guest user). However, he needs to pay for using any service (amount of accessed information or the duration of using other user's connectivity). The payments of Alien and Bill are often based on the amount of information accessed (downloaded) and time usage-based (i.e., proportional to the WiFi connection time to connect to the connectivity shared by other users) [19, 26].

The network operator (Flood Management Center) and the users (Subscribers and Aliens) interact with each other as follows. First, the operator announces the *pricing* and *incentive* mechanism, i.e., the price of flood related information charged to Bill and Alien users; and the percentage of revenue shared with Bill users. Second, each subscriber chooses a membership type for a given time period (e.g., three months), considering his mobility pattern and demand on the flood related information during such a period (e.g., frequently travel to the areas which are vulnerable to flood).

Users can change their membership via our platform. The platform also provides statistics on users' participation. This data is used to adjust the game rules, i.e., the price, the incentive mechanism, and the shared portion with Bill users, to increase the number of users who are willing to share data and their devices to the system. Note however that the analysis of this platform is deferred to future work. The main contribution here is providing a platform to run the flood mitigation processes under a game model thereby flood mitigation tasks can be shared to the community.

In the next section, we describe our proposed overload-aware routing protocol which helps to overcome the bottleneck issues of nodes close to the root in the OEMAN.

5   Overload-Aware Routing Optimization

As mentioned already, OEMAN utilizes a tree-based topology to conveniently establish the Internet connectivity in the shared network. This approach faces load balancing issues where nodes close to the root have to route a much higher traffic volume compared with nodes closer to the tree leaves. In this section, we address the particular issue of imbalanced OEMAN nodes causing bottlenecks or even failure at heavy nodes. We propose a linear optimization model for overload-aware routing that prevents overloading nodes in OEMAN. Our overload-aware strategy makes use of multipath routing to avoid bottlenecks at some nodes. We first provide a brief description of multipath routing. Multipath routing in wireless networks raises the problem of signal interference. Next we describe a solution for avoiding wireless interference, a solution later used in our linear program. Finally we provide the formulation of our linear program, detailing the modeling constraints.

5.1   Multipath Routing

Let the wireless transmission network be modeled by a directed graph $G = (V, E)$, where $i \in V$ is either an AP or a mobile device. Each edge $[e = (i, j)] \in E$ denotes a directed wireless link from node i to node j. Let $R(e)$ be the transmission rate of link e. In Fig. 3a, using a shortest path routing, flows from MN4, MN5 and MN9 to AP are routed as follows: MN4→MN1→AP,

MN5→MN1→AP and MN9→MN5→MN1→AP. We observe that MN1 is overloaded because of the flows sent on the outgoing link of MN1 → AP. In
order to solve this problem, we consider a multipath routing based on the
N-shortest paths between N source nodes and a destination node as illustrated in Fig. 3b. Here, to avoid an overload at MN1, the paths from MN5 and
MN9 to AP are changed as follows: MN5→MN6→MN2→AP and
MN9→MN8→MN7→MN3→AP.

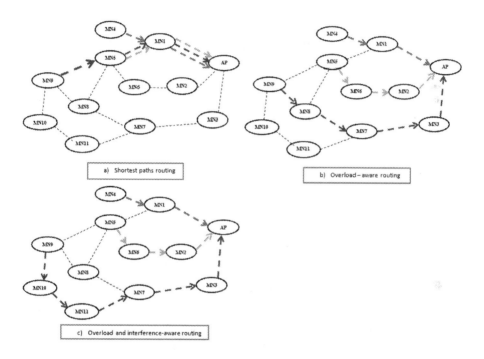

Fig. 3. Illustration of a multipath routing protocol

In Fig. 3b, supposed that MN5 is within the interference range of MN8. Then,
there is a potential conflict between the outgoing link of MN9 to MN8 and the
outgoing link of MN5 to MN6 in transmitting traffic. Hence, in this case, while
MN9 is transmitting traffic to MN8, MN5 should not be transmitting to avoid
collision. To address this potential conflict, the path from MN9 to AP can be
changed as shown in Fig. 3c (MN9→MN10→MN11→MN7→MN3→AP).

5.2   Avoiding Wireless Interference

For unicast transmission scheduling in wireless network, a transmission from a
node i to node j is successful if node j is within communication range of node
i and any node k within interference range of node j is not transmitting. Let
H be a unicast conflict graph whose nodes represent links of a communication

network G. Each node $n_e^i \in H$ is a unicast transmission at node $i \in V$ on outgoing link $e \in E$. Two nodes of the conflict graph H are connected by an edge if the corresponding links cannot be active simultaneously. Let C be a clique that comprises a subset of unicast nodes in H and let C_{max} be a maximal clique in H. Then in each clique C, the wireless unicast transmissions mutually conflict together, at any time only one transmission can be active. The constraint for the unicast transmission scheduling is then modeled as in [25] as followed:

$$\sum_{n_i^e \in C_{max}} \frac{u_{i(e)}}{R(e)} \le 1 \quad \forall \quad C_{max} \ in \ H \tag{1}$$

where $u_{i(e)}/R(e)$ is the fraction of time that unicast node n_e^i is active. For all cliques in H, we only consider the constraints related to the maximal cliques for the unicast transmission scheduling because the constraints related to cliques which are covered by maximal cliques are supernumerary.

Interference-aware routing is a routing which avoids interfering transmissions. To prevent wireless interferences in our multipath routing, relation (1) is included as a constraint in our optimization model such to obtain interference-aware routings.

5.3 Linear Optimization Formulation

Let $E_{in}(i)$ and $E_{out}(i)$ be respectively the sets of incoming links and outgoing links at node $i \in G$. Let X represent the set of demands (each demand is a communication from a node i to a node j). Each demand $x \in X$ has source node $s(x)$, destination node $d(x)$ and its traffic value sent from the source node to the destination node is denoted by $t(x)$. Let θ be defined as a maximum multiplier so that all routing demands with their traffic value multiplied by θ can be feasibly routed in the network. For each demand x, P_x denotes the set of possible routing paths from source node $s(x)$ to destination node $d(x)$. Routing variable $z_{x(P)}$ of each demand $x \in X$ represents the amount of traffic sent on path P. Let variable $u_{(i(e))}$ denotes the total amount of traffic which is unicast on link e at node i. Our linear optimization program for the overload-aware routing is as follows:

$$\text{Maximize } \theta$$

$$\text{s.t.} \sum_{P \in P_x} z_{x(P)} = t(x)\theta \quad \forall \quad x \in X \tag{2}$$

$$u_{i(e)} = \sum_{x \in X, s(x)=i} \sum_{P \in P_x, e \in P} z_{x(P)}$$
$$+ \sum_{e' \in E_{in}(i)} \sum_{x \in X} \sum_{P \in P_x, e' \in P} z_{x(P)} \forall \quad e \in E_{out}(i), i \in V \tag{3}$$

$$u_{i(e)} - q_{i(e)} \le B_i \quad \forall \quad e \in E_{out}(i), i \in V \tag{4}$$

$$\sum_{n_i^e \in C_{max}} \frac{u_{i(e)}}{R(e)} \leq 1 \quad \forall \quad C_{max} \ in \ H \tag{5}$$

Constraint (2) states that the total amount of traffic $z_{x(P)}$ of each demand $x \in X$ that is sent from the source node $s(x)$ to the destination node $d(x)$ among the possible paths must be equal to the traffic value of the demand $t(x)$ multiplied by θ [24].

Constraint (3), modified from [24], states that the total amount of traffic $u_{i(e)}$ which is unicast on link e at node i consists of the amount of original traffic of node i that is sent out on link e and the amount of unicast transit traffic that goes out of node i on link e whose previous-links are a set of $e' \in E_{in}(i)$.

Constraint (4), where $q_{i(e)}$ denotes sending capacity of node i on link e, states that for any node in the network, the load value $(u_{i(e)} - q_{i(e)})$ of that node must be lower than or equal to its buffer size B_i in order for it to avoid overload while transmitting traffic.

Constraint (5), discussed in Sect. 5.2, is related to the unicast transmission scheduling for each maximal clique C_{max} in the unicast conflict graph H.

The above overload-aware routing model satisfies the demand through paths that optimize wireless network throughput, those paths are not necessary the shortest paths. The selected paths avoid some nodes with heavy load to achieve a load balancing among VAPs in OEMAN. In addition, considering wireless interference, this model finds paths without collisions. Our formulation can be used for computing the throughput on any OEMAN. Throughput results are computed and shown in the next section.

6   Prototyping and Evaluation

In this section we present our prototype for the urban flood data sharing platform and conduct a preliminary survey to analyze the users' willingness to share their data. After that we thoroughly present the evaluation of the proposed overload-aware routing (OAR) protocol applied to the OEMAN scheme.

6.1   Prototype and Users' Willingness on Data Sharing

We have developed a prototype for the proposed urban flood mitigation platform which consists of two main parts: (i) the front-end ones consist of mobile applications (running on both the IOS and Android mobile devices) that allow users to collect and share flood related data to the system, (ii) the back-end components consist of database systems and data analytics that pre-process the data (e.g., for data cleaning) and analyze the reported data to provide useful information about floods to users via mobile apps mentioned in (i). Figure 4 shows a screenshot on which a user can collect and share the flood related data including *location mapped on the city map, the estimated time and the user-estimated duration of the flood* to the analysis system. Consequently, users will receive updated flood information displayed on a city map from our analysis system as depicted in Fig. 5.

Using this platform, we conducted a survey to collect users' opinions on the usefulness of the system and users' expected reward when sharing the data. We have received more than 170 answers for two days opening of this survey. The

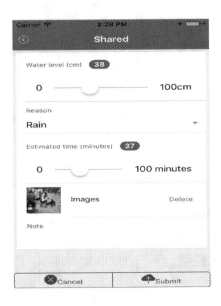

Fig. 4. A user shares the flood data he or she captures to the system

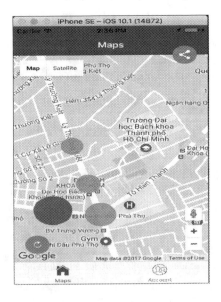

Fig. 5. A city map displays the updated information of a flood

survey results reveal that 92% of users answer that the proposed platform is useful for urban flood mitigation. It is interesting that 60% of users are willing to share data just for contribution to the flood mitigation process and 9% of the users expect to get reputation (from community) making a summation of around 70% users are willing to share data for free. Other 20% of users are expected to get rewards by using services for free of charge, while only 11% of users are expected to get money back upon the shared data.

6.2   Effectiveness of the Proposed OAR Method

In order to evaluate the impact of the OAR in terms of throughput and E2E delays, it is compared with a shortest path routing (SPR) using an implementation in Matlab.

A 32-node OEMAN topology which includes 4 APs (using the case of multiple surviving APs) and 28 mobile nodes has been generated randomly in a flat area of $400\,\mathrm{m} \times 400\,\mathrm{m}$, see Fig. 6. The transmission rate $R(e)$ at each link is 12 Mbps while the communication range is 94.4 m, based on the 802.11a standard as shown in Table 1. The interference range of each node is chosen to be twice the communication range (188.8 m), so the interference range is much larger than the communication range [27].

Table 1. IEEE 802.11a standard.

Transmission rate (Mbps)	Communication range (m)
6	100
12	94.4
18	84.1
24	75.0
36	63.0
48	50.1
54	39.8

Let $N_{t(x)} = \sum_{x \in X} t(x)$ denotes the total traffic value. The source and the destination node of each unit (1 Mb) of the total traffic value is chosen randomly. It is assumed that the buffer size B_i of each node i is 4 Mb while the sending capacity $q_{i(e)}$ of each node is assumed to be 6 Mbps. To demonstrate the benefits of OAR, we try to have B_i and $q_{i(e)}$ so that $B_i < q_{i(e)} < R(e)$ to obtain cases where some VAPs are overloaded. For our multipath routing method, P_x (the set of potential routing paths for demand x) consists of the five shortest paths between s_x and d_x. Furthermore we suppose that the delivery probability of all links in the network is ideal.

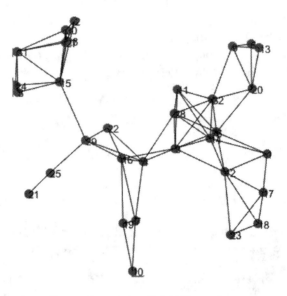

Fig. 6. A 32-node OEMAN topology.

Results of our simulations are presented in Fig. 7. This figure compares the total end-to-end throughput of both the overload-aware routing scheme and the default shortest path routing scheme. We observe that the overload-aware routing achieves a greater end-to-end throughput compared to the default shortest path routing without load balancing. The reason is because OAR finds the paths that have fewer overloaded VAPs and wireless interference. OAR outperforms SPR for all total traffic values. OAR enhances the throughput of OEMAN up to about 58.799% for a total traffic demand of 50 Mb. Even with the high total traffic value of 400 Mb, the throughput of OAR still increases by about 22.188%.

Figure 8 shows the throughput gain of the overload-aware routing scenario as inferred from Fig. 7. The throughput gain is the ratio of the throughput obtained by the overload-aware routing approach to the throughput obtained by the default shortest path routing approach:

$$G_{throughput\_gain} = \frac{N_{z(x)\_OAR}}{N_{z(x)\_SPR}} \tag{6}$$

where $N_{z(x)\_OAR}$ and $N_{z(x)\_SPR}$ are the total end-to-end throughput of all examined demands of OAR and SPR, respectively. For example, for a total traffic demand of 10 Mb, the total end-to-end throughput of OAR is equal to 5.947 Mb and the total end-to-end throughput of SPR is equal to 4.613 Mb. The throughput gain of this case $G_{throughput\_gain} = \frac{N_{z(x)\_OAR}}{N_{z(x)\_SPR}} = \frac{5.947}{4.613} = 1.289$

Figure 8 shows the throughput gain of OAR for different traffic levels. It starts with 1.289 for a total traffic demand of 10 Mb, peaks at 1.588 for 50 Mb, and then drops to 1.265 for a total traffic demand of 100 Mb. First throughput increases with the increase of the demand, and then the throughput gain

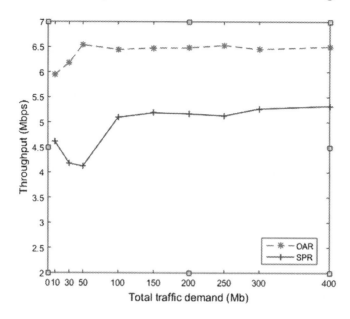

Fig. 7. The effectiveness of the proposed OAR in terms of Throughput.

decreases because it becomes more difficult to find paths without wireless interference and overloaded VAPs. After the total traffic demand of 100 Mb, there are
only slight fluctuations, to stabilize at 1.225 for 300 Mb and 400 Mb. It means
there is no significant increase of the throughput gain when there are so many
demands required at the same time.

Figure 9 shows the total time delay of OAR and SPR. The total time delay
T_{delay_time} is defined as the ratio of the considered total traffic value $N_{t(x)}$ over
the total end-to-end throughput $N_{z(x)}$ corresponding to the traffic value $N_{t(x)}$
as shown in Fig. 7:

$$T_{delay\_time} = \frac{1(s) * N_{t(x)}}{N_{z(x)}} \tag{7}$$

Equation (7) shows, as expected, that the total time delay decreases as the
throughput increases. As shown in Fig. 7, the throughput of OAR is always
greater than that of SPR, consequently the total time delay of OAR is lower
than the one of SPR as shown in Fig. 9. This figure also shows that as total
traffic demand increases, the deviation of time delay between OAR and SPR
also increases. For instance, for a total traffic of 50 Mb, the total end-to-end
throughput in the case of SPR and OAR are 4.119 Mb and 6.540 Mb respectively.
The total time delay of SPR for 50 Mb is $T_{delay\_time} = \frac{1(s)*N_{t(x)}}{N_{z(x)}} = \frac{1(s)*50}{4.119} =$
12.139(s) and similarly the total time delay of OAR for 50 Mb is just 7.645 s.
The time gain is defined as the ratio between the time of transmission required
without load balancing and the one with load balancing. Therefore, for 50 Mb,
the time gain is equal to 1.588. Even in a worst case scenario, with a large

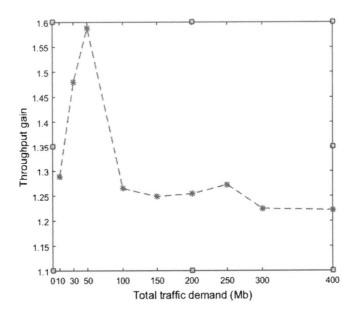

Fig. 8. Throughput gain of the proposed method.

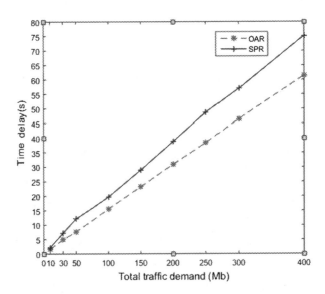

Fig. 9. The effectiveness of the proposed OAR in terms of time delay.

number of victims in a disaster area generating 400 Mb at the same time, it just takes about 61.505 s for the OAR strategy to access the Internet but it will take up to 75.151 s for the SPR strategy, for a time gain of 1.222 in this case.

7 Conclusion and Future Work

Natural disasters may seriously degrade communication infrastructures, leaving victims enable to update the communities outside the disaster area including rescuers and disaster management officers about their situation. This paper proposes an approach to enhance the multihop wireless access networks for flood mitigation crowd-sourcing systems. Concretely, a mobile platform for urban flood data crowd-sourcing and an incentive model for sharing data and communications means (i.e., Wifi cards on individual devices via wireless virtualization mechanism on OEMAN) leveraging game theories has been presented. A prototype for the proposed platform has been developed and preliminary surveys on users' willingness to share data have been conducted. The results reveal that most of users are willing to share urban flood data without expecting to get the monetary reward.

The inherent drawbacks in traffic load imbalance in OEMAN due to its tree-based approach has been thoroughly analyzed and resolved by proposing an optimization model for overload-aware routing (OAR). This approach helps to avoid routing traffic through overloaded VAPs, hence achieves a better load balancing among available VAPs in the network. Evaluations implemented in Matlab confirm the performance of the proposed OAR, in terms of overall throughput and E2E delays, in comparison with the default shortest path routing protocols.

However, to further extend the scope of our analysis, we need to consider in the future work other parameters such packet loss and reduction of load on each node. In addition, to increase the scalability of OEMAN, we are considering approaches based on network coding [20] and distributed consensus techniques [28]. These approaches can address some issues of performance bottlenecks, improve the network capacity and reliability in OEMAN. These should help disaster victims to get an easier access to Internet for emergency relief in regions impacted by a catastrophic disaster. Other issues in OEMAN such as handling mobility, power consumption, node failure, multiple surviving APs will also be considered.

Moreover, the crowd sourcing incentive model needs further verification using real data from real floods. We are planning to deploy our proposed flood data collection and sharing platform combined with the OEMAN access networks in Ho Chi Minh city to collect real data for further analytics for flood prediction and mitigation strategies.

Acknowledgement. This research is funded by Vietnam National University Ho Chi Minh City (VNU-HCM) under grant **number C2017-20-16**.

Appendix A Overload Detection in OEMAN

This section describes functionalities in OEMAN to handle overloaded nodes. Some of these functions have been modified to adapt them to our optimization model. Let L_i denotes the load of node $i \in V$. The load L_i is defined as the total amount of traffic $u_{(i(e))}$ which is unicast on link e at node i minus the sending capacity $q_{i(e)}$ of node i on link e:

$$L_i = u_{(i(e))} - q_{i(e)} \tag{8}$$

Overload is detected at a node i when $L_i > B_i$, where B_i is the buffer size of node i. Overload detection is illustrated in Fig. 10. In this example, PC_1 is the VAP of PC_3 and PC_4, where PC is personal computer. PC_1 examines its load [Algorithm 1: line 7–9] by

$$L_{PC_1} = u_{PC_1(e(PC_1,AP))} - q_{PC_1(e(PC_1,AP))} \tag{9}$$

where $q_{PC_1(e(PC_1,AP))}$ is the sending capacity of node PC_1 on the link PC_1 to AP and $u_{PC_1(e(PC_1,AP))}$ is the total amount of traffic from PC_3 and PC_4 which is unicast on link $e(PC_1, AP)$ at PC_1. PC_1 detects that it is overloaded ($L_{PC_1} > B_{PC_1}$) [Algorithm 1: line 7] since the traffic from two sub-trees $t1$ and $t2$ rooted by PC_3 and PC_4 is greater than the buffer capacity of node PC_1. In that case, PC_1 asks for help from other PCs for sharing load. Consequently, PC_1 will broadcast its overload messages to all PCs managed by it, namely PC_3 and PC_4. Immediately, PC_4 moves to another VAP (PC_2) [Algorithm 1: line 10–11] to transmit its traffic to AP. Examining wireless interference under the protocol model of interference [29] is also performed before doing a handover to PC_2. Consequently, load balancing among VAPs is achieved.

Algorithm 1 describes the computing of load at a VAP and the execution of the handovers in the case of the overloaded VAP.

Algorithm 1. Load-sharing

1: **procedure** LOAD_SHARING(*double* θ) ▷ At the serving node (e.g. PC_1 in Fig. 10)
2:     *List* $L = sort(thisNode.childrenList)$;
3:     *double* $buff = thisNode.buffersize$;
4:     *double* $q = thisNode.SENDING\_CAPACITY()$; ▷ Call Algorithm 3, where q is the sending capacity of serving node
5:     *double* $p = 0, load = 0$;   ▷ p is the traffic from sub-trees (e.g. two sub-trees $t1$ and $t2$ in Fig. 10)
6:     *int* $i = 0; load = L.element(i).traffic - q$;
7:     **while** $load <= \theta.buff$ **do**   ▷ the condition to dectect the overload
8:         $i + +$;
9:         $load+ = L.element(i).traffic$;   ▷ Compute the load
10:    **for** $j = i$ to $L.length$ **do**   ▷ Asking appropriate nodes to handover
11:        $HANDOVER(L.element(j))$;   ▷ Call Algorithm 2

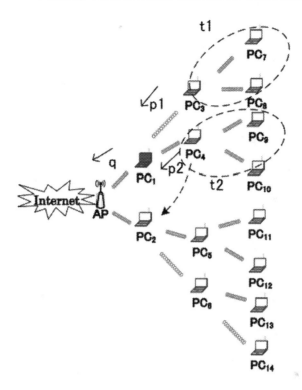

Fig. 10. A node with heavy load in a tree-based network

Algorithm 2 is the handover function which considers the conditions of doing a handover of any client node.

Algorithm 2. Handover

1: **procedure** HANDOVER(*Node j*)   ▷ At a client node (e.g. *PC₄* in Fig. 10)
2:     *Select the neighborNode k;*
3:     *From j.neighborList();*
4:     *Where k.LOAD() >= j.traffic;*   ▷ Call Algorithm 4
5:     *And k.hopcount is the lowest in j.neighborList();*
6:     *And k.collisions is minimum;*
7:     *Handover to node k;*

Algorithm 3 is the function returning the value of sending capacity of examining node.

Algorithm 3. Sending Capacity

1: **procedure** SENDING_CAPACITY()
2:     *return thisNode.sendingRate;*   ▷ Assume that this value can be estimated

Algorithm 4 is the function returning the value of a VAP which can be consider for serving handover of a client node.

Algorithm 4. Load

1: **procedure** LOAD()
2:     $double \ q = thisNode.SENDING\_CAPACITY()$;   ▷ Call Algorithm 3
3:     $double \ p = 0; double \ buff = thisNode.buffersize$;
4:     $List \ L = thisNode.childrenList$;
5:     **for** $i = 0 \ to \ (L.length - 1)$ **do**
6:         $p+ = \ L.element(i).traffic$; ▷ Compute the total amount of transit traffic of examining node
7:         **if** $((p - q) \ <= 0)$ **then** ▷ Return the value of VAP which can be consider for serving handover of a client node
8:             $return \ ((q - p) + buff)$;
9:         **else**
10:            **if** $((p - q) >= buff)$ **then**
11:                $return \ 0$;
12:            **else**
13:                $return \ (buff - (p - q))$;

References

1. Quang, T.M., Kien, N., Yamada, S.: DRANs: resilient disaster recovery access networks. In: The 1st IEEE International Workshop on Future Internet Technologies (IWFIT 2013), in conjunction with IEEE COMPSAC 2013, pp. 754–759, July 2013
2. Ogura, J., Park, M., Wakatsuki, Y., Sanchez, R.: Japan earthquakes: dozens killed; race against the clock to find survivors. CNN, April 2016. http://edition.cnn.com/2016/04/16/asia/japanearthquake/
3. McKirdy, E., Singh, H.S.: Deadly earthquake shakes part of Northeastern India. CNN, January 2016. http://edition.cnn.com/2016/01/03/asia/india-earthquake/
4. Hanna, J.: 43 dead after typhoon koppu hits Philippines. CNN, October 2015. http://edition.cnn.com/2015/10/22/asia/typhoonkoppu-lando-philippines/
5. Quang, T.M., Nguyen, K., Cristian, B., Yamada, S.: On-the-fly establishment of multihop wireless access networks for disaster recovery. IEEE Commun. Mag. **52**(10), 60–66 (2014)
6. Quang, T.M., Shibata, Y., Cristian, B., Yamada, S.: On-site configuration of disaster recovery access networks made easy. Ad Hoc Netw. **40**, 46–60 (2016). Elsevier
7. Chandra, R., Bahl, P.: Multinet: connecting to multiple IEEE 802.11 network using a single wireless card. IEEE INFOCOM Hong Kong, pp. 882–893, March 2004
8. Ho Chi Minh City: Steering center of the urban flood control program, February 2017. http://www.ttcn.hochiminhcity.gov.vn/
9. Plate, E.J.: Flood risk and flood management. J. Hydrol. **267**, 2–11 (2002)
10. van der Kooij, M.: Flood monitoring and disaster management response. GEOconnexion Int. Mag. 26–28 (2013)

11. de Brito Moreira, R., Degrossi, L.C., De Albuquerque, J.P.: An experimental evaluation of a crowdsourcing-based approach for flood risk management. In: 12th Workshop on Experimental Software Engineering (ESELAW), At Lima, Peru, pp. 1–11, April 2015

12. Do, H.N., Vo, M.T., Tran, V.S., Tan, P.V., Trinh, C.V.: An early flood detection system using mobile networks. In: International Conference on Advanced Technologies for Communications (ATC), Ho Chi Minh City, Vietnam, pp. 599–603, October 2015

13. Gomes, J.L., Jesus, G., Rogeiro, J., Oliveira, A., Tavares da Costa, R., Fortunato, A.B.: Molines - towards a responsive web platform for flood forecasting and risk mitigation. In: Federated Conference on Computer Science and Information Systems (FedCSIS), pp. 1171–1176 (2015)

14. Amrani, C.E., Rochon, G.L., El-Ghazawi, T., Altay, G., Rachidi, T.: Development of a real-time urban remote sensing initiative in the Mediterranean region for early warning and mitigation of disasters. In: IEEE International Geoscience and Remote Sensing Symposium (IGARSS), pp. 2782–2785 (2012)

15. Serpico, S., Dellepiane, S., Boni, G., Moser, G., Angiati, E., Rudari, R.: Information extraction from remote sensing images for flood monitoring and damage evaluation. Proc. IEEE 100(10), 2946–2970 (2012)

16. Mandal, S., Saha, D., Banerjee, T.: A neural network based prediction model for flood in a disaster management system with sensor networks. In: International Conference on Intelligent Sensing and Information Processing, pp. 78–82 (2005)

17. Perez, P., Holderness du Chemin, T., Turpin, E., Clarke, R.: Citizen-driven flood mapping in Jakarta: a self-organising socio-technical system. In: IEEE International Conference on Self-adaptive and Self-organizing Systems Workshops (SASOW), pp. 174–178 (2015)

18. Funayama, T., Yamamoto, Y., Tomita, M., Uchida, O., Kajita, Y.: Disaster mitigation support system using Twitter and GIS. In: International Conference on ICT and Knowledge Engineering, pp. 18–23 (2014)

19. Ma, Q., Gao, L., Liu, Y.F., Huang, J.: A game-theoretic analysis of user behaviors in crowdsourced wireless community networks. In: 13th International Symposium on Modeling and Optimization in Mobile, Ad Hoc, and Wireless Networks (WiOpt), pp. 355–362 (2015)

20. Ahlswede, R., Cai, N., Li, S.Y., Yeung, R.W.: Network information flow. IEEE Trans. Inf. Theor. 46(4), 1204–1216 (2000)

21. Katti, S., Rahul, H., Hu, W., Katabi, D., Médard, M., Crowcroft, J.: XORs in the air: practical wireless network coding. SIGCOMM Comput. Commun. Rev. 36(4), 243–254 (2006). http://doi.acm.org/10.1145/1151659.1159942

22. Fan, K., Wei, X., Long, D.: A load-balanced route selection for network coding in wireless mesh networks. In: Proceedings of IEEE International Conference on Communications, ICC 2009, Dresden, Germany, pp. 1–6, 14–18 June 2009. http://dx.doi.org/10.1109/ICC.2009.5198624

23. Maxemchuk, N.F.: Dispersity routing. In: International Conference on Communications (ICC 75), pp. 41.10–41.13, San Francisco, CA, June 1975

24. Sengupta, S., Rayanchu, S., Banerjee, S.: Network coding-aware routing in wireless networks. IEEE/ACM Trans. Netw. 18(4), 1158–1170 (2010)

25. Jain, K., Padhye, J., Padmanabhan, V.N., Qiu, L.: Impact of interference on multihop wireless network performance. In: Proceedings of ACM MobiCom, pp. 66–80, September 2003

26. Fon website, February 2017. https://corp.fon.com/en

27. Zhang, Y., Chen, H.-H., Guizani, M.: Cooperative Wireless Communications, 1st edn. CRC Press, Boca Raton (2009)
28. Olfati-Saber, R., Fax, J.A., Murray, R.M.: Consensus and cooperation in networked multi-agent systems. Proc. IEEE **95**(1), 215–233 (2007)
29. Gupta, P., Kumar, P.R.: The capacity of wireless networks. IEEE Trans. Inf. Theory **46**(2), 388–404 (2000)

Assessment of Aviation Security Risk Management for Airline Turnaround Processes

Raimundas Matulevičius[1], Alex Norta[2], Chibuzor Udokwu[2(✉)],
and Rein Nõukas[2]

[1] Institute of Computer Science, University of Tartu,
J. Liivi 2, 50409 Tartu, Estonia
rma@ut.ee
[2] Department of Software Science, Tallinn University of Technology,
Akadeemia tee 15A, 12816 Tallinn, Estonia
alex.norta.phd@ieee.org, chibuzor.udokwu@ttu.ee, rein.noukas@gmail.com

Abstract. Security in the aircraft business attracts heightened attention because of the expansion of differing cyber attacks, many being driven by technology innovation. Continuous research does not consider the sociotechnical essence of security in basic areas, for example, carrier turnaround systems. To cut time and costs, the latter comprises several companies for ticket- and luggage management, maintenance checks, cleaning, passenger transportation, re-fueling, and so on. The carrier business has embraced broadly data innovation for guaranteeing that aircrafts are in a state to take off again as fast as would be prudent. Progressively, this prompts the development of a virtual enterprise that utilizes data advances to consistently coordinate individual airline-turnaround processes into a single structure. The subsequent sociotechnical security risk management issues are not clearly understood and require further examination. This paper fills the gap with an assessment about the application of a security risk management strategy to identify business assets for a more profound risk mitigation analyses. The result of this paper provides knowledge about the usefulness of existing security risk management approaches.

Keywords: Security · Risk analysis · Airline turnaround
Virtual organization · Decentralization · Composition
Mitigation · Sociotechnical · e-governance · Business process
Cross-organizational

1 Introduction

The airline industry experiences a significant application of technology innovation in all aspects [5]. Numerous novel risk- and security issues are related with civil-aviation communication systems that result in worst cases in catastrophic

A. Hameurlain et al. (Eds.): TLDKS XXXVI, LNCS 10720, pp. 109–141, 2017.
https://doi.org/10.1007/978-3-662-56266-6_6

crashes of aircrafts. Other critical security issues are identified with communication, for example, a deliberate jamming of Automatic Dependent Surveillance-Broadcast (ADS-B) systems [11]. ADS-B is a surveillance system in which an aircraft locates its position via satellite navigation and periodically communicates that position for tracking by air traffic control ground stations. This confirms that the aviation industry is rapidly turning into a cyber-physical system (CPS) [24] that poses more risks and security issues. Briefly, a CPS [4] is a system composed of physical components that are controlled and monitored by computer algorithms.

While the underlying way to deal with examining airport-related security is largely technical, recent work recognizes this is a socio-technical systems [12] matter. A sociotechnical system is characterized by a complex organizational work design where people require sophisticated technology tools to solve problems at the work place. In [14], the authors recognize for the first time that airports are social technical in nature by using extended use-case diagrams and storyboard representations of use cases to discover stakeholder requirements such as security for the development of an airport operating system. Security in connection with information is explicitly investigation in [8] that compares practitioner-oriented risk management methods and several academic security modeling frameworks to develop an ontology, or domain model, of information system security risk management. More recently, in [15] the authors investigate requirements evolution in the context of the SecureChange[1] EU-venture in which the business case is drawn from the Air Traffic Management (ATM) area. While safety and security specialists are part of the focus groups, the contextual analysis do not explicitly zoom into security specifics in their study results. Besides, parameter quantifiability and social aspects of security arrangements in [25] examining particularly the costs versus benefit trade-off in alternative airport security policies constellations pertaining to, e.g., passengers, items such as baggage, and so on.

Literature shows security based research for aviation industry is an important research interest. However, the security topics are not broad and also do not consider that modern information technology enables ad-hoc and process-aware cross-organizational collaborations [6,7,10,20–22] that improves the reduction of time and costs of airline management while improving quality service delivery. Such novel methods for airline management systems additionally prompt unusual security risk issues for which the mitigation techniques are not clearly defined. This paper fills the gap with an evaluation that answers the research question of *how to use information system security risk management* (ISSRM) *in cross-organizational collaborations* (COC) *of the airline industry to achieve a desired level of security.* For establishing a separation of concerns, we deduce the following sub-questions: What are relevant assets in airline COC that need to be secured? What security risks threaten COC systems? What are security risk mitigation strategies in airline COC?

---

[1] http://www.securechange.eu/.

This paper is an extension of the conference paper [16]. In this work, we also illustrate how identified security risks and security counter measures could be estimated using process simulation.

The rest of the paper is structured as follows. Section 2 gives background information that is necessary for being able to follow the rest of the paper. Section 3 gives essential properties that characterize agent negotiation. Next, Sect. 4 finds security risks that threaten airline turnaround systems. Section 5 gives mitigation strategies to protect from security risks. Section 6 provides a risk assessment with focus on control selection to prioritize risk mitigation as a result of limited resources. Section 7 shows a security trade-off analysis process and finally, Sect. 8 concludes with important lessons learned and future work directions.

2 Background

We provide additional information that is relevant for the remaining part of this paper. First, Sect. 2.1 is derived from a running case that results from studying COC in an airline-turnaround scenario. Secondly, Sect. 2.2 shows parts of a security risk management framework that is applicable for studying our running case.

2.1 Running Airline-Turnaround Case

We use the business process model notation BPMN 2.0 [18] for showing the running case in Fig. 1. BPMN is a widely accepted standard for modeling processes and provides a graphical notation showing process steps as boxes of various kinds and their order by joining them with arrows.

The simplified airline turnaround process in Fig. 1 is taken from a larger model in [19] and shows three pools for ground services, passenger management and gate agent in that order. The ground services begins with a start signal event to start after-flight services, followed by an AND-split. The top parallel branch has a catching intermediary start signal event for when all passengers have de-boarded, followed by yet another AND-split parallel gateway for cleaning, restocking aircraft, and fueling after a start message event indicates the receipt of a fuel slip. The task for restocking the aircraft requires a data object comprising passenger information, e.g., about specific dietary needs. After an AND-join, an intermediate signal event informs that boarding is now allowed. The bottom branch of the initial AND-split begins with a task for offloading cargo and luggage, followed by an AND-split with parallel branches and respective intermediate message event nodes. The top parallel branch waits via a catching intermediate event for a message from an adjacent process that indicates a cargo assignment and the second parallel branch likewise needs to wait until catching the message that the luggage receipt exists. After the AND-join, cargo- and luggage-loading commences, culminating into another AND-join before an end-signal event terminates the process for ground services.

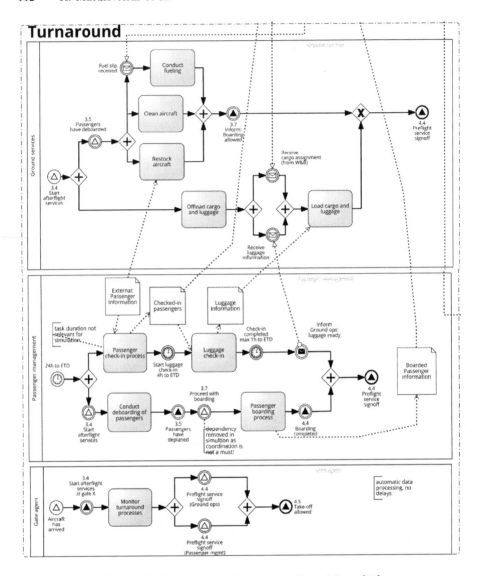

Fig. 1. Airline turnaround process, adapted from [19].

The middle pool of Fig. 1 for passenger management commences with a start timer event, namely, 24 h before the estimated time of departure (ETD). The latter is followed by an AND-split with the top parallel branch starting a sequence with a task for passenger check-ins. The latter takes a data object as input comprising external passenger information and the task also contributes to a data object about checked in passengers. Next in the sequence follows an intermediate timer event to wait until 4 h before ETD for luggage check-in. The actual task for luggage check-in uses the data object about checked in passengers and produces new facts for a data object abut luggage information. After the completed check-in and given there is one hour left until ETD, an intermediate message event sends information for the ground operations about the luggage being ready to the swimlane for ground services before the AND-join. The bottom parallel branch starts with an intermediate signal event to start after-flight services, followed by a task for conducting the de-boarding of the passenger from the landed airplane. After that, one intermediate signal event indicates all passengers have deplaned, followed by yet another intermediate signal event in the sequence stating that boarding may proceed. The subsequent task for the actual boarding process affects the data object for boarded passenger information. Following the final intermediate signal event about boarding having completed, the AND-join leads to the end signal event for signing off the preflight service.

The final pool in Fig. 1 for the gate agent commences with the start signal event that the aircraft has arrived, followed by the intermediate signal event for starting the afterflight services at a specific gate. Furthermore, the gate agent monitors via a task the turnaround process before an AND-split where in parallel two intermediate signal events indicate a preflight service sign-off for ground operation and for passenger management respectively. The subsequent AND-join leads to the end signal event for allowing an airplane takeoff.

2.2 Security Risk Management Domain Model

The ISSRM domain model [8,17] consists of asset-related, risk-related and risk treatment-related concepts. In Fig. 2, this domain model is presented as a UML class diagram for which we briefly present the concepts below.

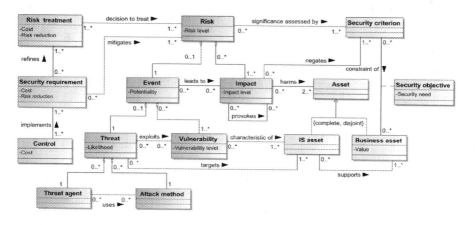

Fig. 2. The ISSRM domain model, adapted from [8,17].

Asset-related concepts describe organizational assets that need to be protect and the criteria for determining a satisfactory level of security. An *asset* is anything that is valuable to the organization and plays a role achieving the organization's objectives. Assets can be grouped into business assets and information-system assets. A *business asset* describes the information, processes, capabilities and skills essential to the business and its core mission. The *value* metric is used to identify the security need of each asset in terms of confidentiality, integrity and availability (see below). An *IS asset* that we later depict as a *system asset* too, is a part of an information system that is valuable to an organization and supports business assets.

A *security criterion* indicates the security need as a property on business assets. The security criterion describes the security needs that are usually expressed as confidentiality, integrity and availability of business assets. A metric to assess a *security need* expresses the importance of security criterion with respect to business asset.

Risk-related concepts introduce risk definitions and its components. A *risk* is the combination of a threat with one or more vulnerabilities leading to a negative impact harming at least two or more assets. An *impact* is the potential negative consequence of a risk that negates the security criterion defined for business assets in order to harm these assets when a threat (or an event) is accomplished. A risk *event* is an aggregation of threat and one or more vulnerabilities. A *vulnerability* is the characteristic of an IS asset or group of IS assets that expose a weakness or flaw in terms of security. A *threat* is an incident initiated by a threat agent using an attack method to target one or more IS assets by exploiting their vulnerabilities. A *threat agent* is an agent who has means to intentionally harm IS assets. A threat agent triggers a threat and, thus, is the source of a risk. The threat agent is characterized by expertise, his available resources, and motivation. An *attack method* describes a standard means by which a threat agent executes a threat.

Risk is estimated using a *Risk level* metric. The risk level depends on the event *Potentiality* and the *Impact level*. An event's *Potentiality* depends on the threat *Likelihood* and *Vulnerability level*. It is necessary to note that a threat agent and attack method do not have their own metrics representing their level. Some characteristics of threat agents and attack methods can be identified independently, e.g., an agent's motivation and experience. Still, they can also be used as indicators to estimate the likelihood of a threat.

Risk treatment-related concepts describe concepts to treat risk. A *risk treatment-decision* is a decision to treat an identified risk. A treatment satisfies a security need, expressed in generic and functional terms and are refined to security requirements. There are four categories of risk treatment decisions possible – risk avoidance, risk reduction, risk transfer, and risk retention. A *security requirement* is a condition over the phenomena of the environment that we wish to make true by installing the information system, in order to mitigate risks. Finally, a *control* is a designed means to improve the security by implementing security requirements. Risk treatment and security requirements are estimated in terms of *Risk reduction* performed and *Cost* incurred; Controls – in terms of *Cost*.

Process. In [17], security risk management is described as an analysis result of the security- and security risk management standards. The process in Fig. 3 begins with (i) a study of the organization's context and the identification of its assets. Then, one needs to determine the (ii) security objectives in terms of confidentiality, integrity and availability of the business assets. The next step of the process is (iii) risk analysis where security risks are elicited and assessed. Once risk assessment is finished, decisions about (iv) risk treatment are taken. Step (v) is the elicitation of security requirements to mitigate the identified risks. Finally, security requirements are implemented to security controls (vi). The ISSRM process is iterative until reaching an acceptable level for each risk.

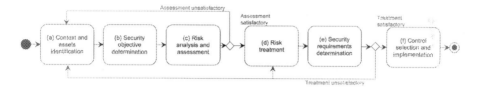

Fig. 3. Process for security risk management, adapted from [8, 17].

3 Asset Identification and Security Objective Determination

To analyze security issues in cross-organizational business processes, the first step is to identify assets involved in the airline turnaround collaboration in Fig. 1 and determine their security objectives. These two tasks correspond to the two first steps illustrated in Fig. 2. Our goal is to identify assets that are involved in the collaboration between the airline and service providers. Consequently, Sect. 3.1 focuses on the airline turnaround day of operation (DOO). We analyze what IT systems are involved in this turnaround process. Section 3.2 identifies business assets that contain information exchanges between the airline and service providers.

3.1 IT Systems for Airline Turnaround

The airline day of operation shows routine tasks and processes before and after takeoff flights. According to [19], we consider the processes for flight preparation, turnaround and takeoff. Following the depiction in Fig. 1, the process of flight preparation involves gathering and compiling of all flight plans. The latter are documents that describe proposed aircraft flights.

The turnaround phase of operations involves the following set of activities: ground operations, passenger management and gate-agent activities. The ground operations encompass all activities that take place before the passengers start boarding the aircraft. Cargo and luggage offload, aircraft cleaning, restocking of aircraft, re-fueling and loading of cargo and luggage. Passenger management comprises passenger check and luggage check-in activities. The gate agent monitors the ground operations activities and passenger-management activities.

The takeoff activities are the last set of pre-flight activities to be carried out before the actual takeoff of a flight. The activities include reviewing of flight plans, load balancing and the calculation of additionally required fuel. This phase ends with the approval of a flight plan and the request for takeoff clearance from the air-traffic control (ATC).

Out of all the airline DOO processes described, the turnaround phase provides more opportunities for collaborations between the airline and service providers. This is because the activities involved in this phase are resource intensive and are not part of the core competence of the airlines.

From Fig. 1, we derive the following information systems that support collaboration activities in the turnaround process. Passenger management as an IS asset that contains activities, participants, business entities, roles and rules in a passenger-management pool of the turnaround process. Ground operations comprising activities, business entities etc. in the ground operations pool of the turnaround process. The messaging system with rules, protocols and networks that determine how digital information is transmitted between airlines and service providers. For example, the domain-name servers (DNS) and simple mail transfer protocol (SMTP) are networks and protocols that play a role in the delivery of messages from sender to receiver. The passenger check-in process is an IS asset that contains rules, procedures etc. for boarding passengers. Finally, the luggage check-in process as an IS asset that contains rules, procedures etc. on how luggage is checked-in to an aircraft.

3.2   Information Exchange in Collaborating Systems

The knowledge model for airline turnaround is depicted in Fig. 1. It comprises the following roles - passenger management and ground operations. Each set of activities generates specific data objects and can also trigger other sets of activities that we give below.

Passenger Management Process: Different forms of passenger data are generated throughout the passenger-management process. Such data may include names, addresses, phone numbers, next of kin, etc. Other data contained may include frequency of travel, destination and hotel reservations of travelers. The value of this information is significant to the airlines because it contains details that completely describe the customers of the airline. Customers (information) are intangible business assets that must be valued and managed [1], and therefore there is a need to secure customer information as a business asset.

The asset identification starts by identifying information-system assets that supports business assets in the passenger-management processes of the airline-turnaround domain. A process description outlining possible activities involving the business asset is shown in the asset identification Tables 1, 2, 3 and 4 together with the security criteria for each of the business assets identified in the passenger-management processes. Passenger information is contained in the following data objects.

Table 1. Checked-in passenger information asset identification

Business asset	Checked-in passenger information	
IS asset	Passenger check-in process; Passenger management; Passenger; Check-in Personnel	
Process description: how do IS assets *support* business asset(s)	Passenger *physical check-in* process description: • Passenger goes to the check-in personnel • Passenger provides identification document • Personnel verifies provided document • Personnel prints out boarding-pass • Passenger collects boarding pass	Passenger *online check-in* process description: • Passenger visits the online check-in portal • Passenger enters booking number and confirms check-in • Passenger prints boarding-pass
Security criteria	Confidentiality of checked-in passenger data	

The checked-in passenger information of Table 1 comprises data objects that are generated during the check-in activity and contains data about passengers that checked in for the flight. The data may include time of check-in, seat reservations, and other special requests by passengers. The passenger information also contains personal details such as name, passport number, address, contact, next of kin etc. An attacker with access to this information can successfully conduct social-engineering- or phishing attacks on airline passengers.

The check-in activity represents the IS asset that supports the business asset Checked-in passenger information. An attacker can manipulate the passenger check-in process and this may cause blacklisted individuals to be able to board the aircraft.

Table 1 shows details about information system assets that support the business asset of checked-in passenger information. The table also shows two process descriptions involving the business assets, namely physical check-in process and online passenger check-in process. The security criterion for the business asset is also identified as confidentiality of checked-in passenger information. Confidentiality of information is necessary because it is important that data contained in checked-in passenger information are only available for people who should have access to it.

The luggage information of Table 2 shows the data object is created in the luggage check-in activity that starts at the end of passenger check-in activity. The luggage information is transmitted via a messaging system to the ground services to generate cargo assignment information. The data object contains details about baggage carried by different passengers. These may include size, weight and content of passenger baggage.

Table 2. Luggage-information asset identification.

Business asset	Luggage information
IS asset	Luggage check-in process; Messaging system; Passenger management
Process description: how do IS assets *support* business asset(s)	Luggage check-in process description: • Passenger drops the luggage after passenger check-in process • Passenger provides identification document • Personnel measures luggage to confirm if it meets requirement • Personnel records the weight and size • Personnel drops the luggage for loading into aircraft
Security criteria	Confidentiality of luggage information; Integrity of luggage information

The check-in activity and the messaging system represent the IS asset tasks that supports the business asset termed luggage information. An attacker can manipulate the luggage check-in process that may cause luggage with dangerous substances to be loaded to the aircraft.

Table 2 shows details about luggage information business asset. It starts by identifying the IS assets that support the luggage information for the luggage check-in process, the passenger management pool, and the messaging system. The security criteria for the asset are identified as confidentiality of data and integrity of data. This is important because it is necessary that the data contained in luggage information are only available to the right persons and also remain unchanged.

Ground Operations: The following assets are involved in ground operation activities.

The fuel-slip asset is shown in Table 3. After passengers completely de-board the aircraft, a fuel slip is sent via a messaging system to an external provider to start the refueling activity. The fuel slip contains details about the quantity and quality of fuel to be loaded in various fuel tankers of the aircraft. The messaging system represents an IS asset that supports the fuel-slip-business asset.

The quantity of fuel and distribution in the aircraft is very crucial for evenly spreading weight across the aircraft and maintaining proper load balancing. Controlling the latter refers to the location of the center of gravity of an aircraft. This is of primary importance to aircraft stability and determines safety in flight [1].

The data contained in the fuel receipt can be maliciously changed by an attacker, e.g., to cause not enough fuel to be loaded on the aircraft. It is also

Table 3. Fuel-slip asset identification.

Business asset	Fuel slip
IS asset	**Messaging system; Ground operations**
Process description: how do IS assets *support* business asset(s)	Sending fuel slip to service provider process description: • Service provider receives fuel slip • Service provider re-fueling based on data contained in fuel slip
Security criteria	Integrity of fuel slip

possible that an attacker can change the type and quality of fuel on the fuel receipt and this can cause the aircraft to be loaded with wrong fuel. Furthermore, if the wrong quantities of fuel are loaded on different fuel tankers of the aircraft, it can cause the center of gravity of the aircraft to shift beyond allowable limits. This can cause the aircraft to lose stability and spin in midair [1]. Loading an aircraft with the wrong type of fuel results in failure of the engines of the aircraft and can result in air crashes [2].

The fuel slip asset identification in Table 3 shows details of IS assets that support the fuel slip, process description that involves the fuel slip and security criterion for the fuel slip. Two IS assets supporting the fuel slip business asset are messaging system and ground operations pool. The security criterion is identified as integrity of data. It is necessary that data contained in the fuel slip document remains unchanged in the course of airline turnaround.

The cargo assignment in Table 4, commences upon the completion of offloading cargo and luggage. Cargo assignment information is sent via a messaging system to the external provider for commencing the loading of new cargo and luggage into the aircraft. The cargo assignment holds data about weight of baggage, luggage and other check-in cargos. The cargo weights as well as passengers and fuel weights are necessary in maintaining the center of gravity and stability of the aircraft. The messaging system represents the IS asset that supports the business asset cargo assignment.

An attacker can change the values of data contained in the cargo assignment and cause the aircraft to be overloaded beyond an acceptable weight level. This can reduce the efficiency of the aircraft and also reduce the safety margin available if an emergency condition should arise [1]. The reduction in efficiency of the aircraft can result in the following - higher takeoff speed, longer takeoff run, reduced rate and angle of climb, lower maximum altitude, shorter range, reduced cruising speed, reduced maneuverability, higher stalling speed, higher landing speed, longer landing roll[2].

---

[2] Annex of Weighing-systems.com, describing deficiencies of aircrafts as a result of too much weight, http://tinyurl.com/7kborcf.

Table 4. Cargo-assignment asset identification.

Business asset	Cargo assignment
IS asset	Messaging system; Ground operations
Process description: how do IS assets *support* business asset(s)	Sending cargo assignment to service provider process description: • Service provider receives cargo assignment document • Aircraft is loaded with cargoes and luggage based in data contained in cargo assignment document
Security criteria	Integrity of fuel slip

Table 4 shows the IS assets that support the business asset, process description involving the business asset and security criterion for the business asset. The IS assets are messaging system and ground operation pool while the security criterion is integrity of cargo assignment.

4   Security Risk Analysis

We now apply the ISSRM domain model (see, of Fig. 2) to identify security risks in the *passenger management* and *ground operation* processes. This activity corresponds to the third step of the process illustrated in Fig. 3. The activity starts with identifying potential threat agents, their motivation and the resources that they possess to conduct the attack method. Next, we describe a process about how a threat agent is able to carry out an attack method. The risk analysis, then, continues with the identification of the vulnerability (as a characteristic of the IS asset) and impact that describes how security event harms both business and IS asset and how it negates the security criteria. The risk components (such as threat, event, and risk) are defined as the aggregation of the threat agent, attack method, vulnerability and impact, as described in Fig. 2.

In this section, firstly, we detail two risks identified for the *Checked-in passenger information* asset as is defined in Table 1. Next, we overview other security risks pertaining to *luggage information*, *fuel slip*, and *cargo assignment* business assets.

4.1   Analysis of Risks to the *Checked-in Passenger Information* Asset

Two possible attack methods are described for the checked-in passenger information business asset. Each respective attack method has it owns threat agent. Table 5 describes each threat agent, the attack method and risk components for the check-in passenger information.

Table 5. Checked-in passenger information risk- and threat analysis.

	Risk 1	Risk 2
Threat agent	Blacklisted passenger <u>Motivation</u>: need to board the flight <u>Resources</u>: fake ID, money to bribe the check-in personnel <u>Expertise</u>: knowledge of the check-in process	An attacker <u>Motivation</u>: need to board the flight, sabotage the reputation of the airline and cause airline passengers to miss their flights <u>Resources</u>: fake check-in website, passengers data <u>Expertise</u>: knowledge of check-in process, knowledge email phishing attacks
Attack method	Bribes personnel to steal checked-in passenger information Presents fake ID at the check-in desk Gets checked in with fake ID and checked-in passenger information	Attacker sends phishing email to passengers that booked a flight. Passenger enters booking number to the fake check-in website and checks in Passenger prints a fake boarding-pass with flight time changed to few hours ahead of the actual flight time Attacker uses passenger booking number to check-in to the original site and prints boarding-pass Attacker boards the flight with the original boarding pass. Passenger misses flight
Threat	Blacklisted passenger bribes the personnel, presents fake ID, and gets checked-in	Attacker uses phishing email to extract passenger booking number and uses it to check-in to the flight
Vulnerability	Check-in personnel could be bribed	Passenger cant differentiate between original and fake check-in website
Event	Blacklisted passenger presents fake document, bribes personnel and gets checked-in because check-in personnel could be bribed	Attacker uses phishing email to extract passenger booking number and uses it to check-in to the flight because passenger can't differentiate between original and fake check-in website
Impact	Loss of confidentiality of checked-in passenger information Passenger check-in process can no longer be trusted Checked-in passenger information is stolen	Loss of trust in online check-in process Passenger information is stolen Passenger misses flight
Risk	Blacklisted passenger presents fake document, gets checked-in because personnel could be bribed which results to loss of confidentiality of checked-in passenger, loss of trust in check-in process and stolen checked-in passenger information	Attacker uses phishing email to extract passenger booking number and uses it to check-in to the flight because passenger cant differentiate between original and fake check-in website which causes the passengers to miss their flight, their information stolen, resulting to loss of trust in the airline and its online check-in process

The analysis in Table 5 starts by identifying two potential threat agents, namely a blacklisted passenger and an attacker. We assume for the blacklisted passenger, his motivation to carry out an attack on check-in passenger information is the need to board the flight that he is blacklisted for. For a random attacker, his motivation is to sabotage the reputation of the airline by causing the passengers to miss their flights. The two possible attackers have knowledge of how the airline check-in process works. We also assume the attacker who wants to harm the reputation of the airline also needs a fake website to carry out a phishing attack [3] on the passengers of the airline.

Table 5 outlines step-by-step details on how these two attacks are carried out against checked-in passenger information. The first attack method describes a physical passenger check-in process while the second attack description shows an attack on online check-in. The vulnerability in the online check-in process is the passenger ignorance of the genuine check-in website. For the physical check-in process, we assume the check-in personnel is bribed. By successfully taking advantage of these weaknesses, an attacker boards a flight with specific passenger information, or even causes a passenger to miss the flight.

The two attacks in Table 5 harm the airline check-in process. The attacks cause the passengers to loose trust in the passenger check-in process and also the passengers' data are stolen in the course of the attack. Specifically for the second attack, the passenger misses the flight as a result of the attack.

4.2   Analysis of Risks to the *Luggage Information* Asset

Table 6 shows possible attack methods are described for the luggage information business asset. The same possible threat agent - luggage check-in personnel is identified for these two attacks. The Table 6 describes each of the threat agent, the attack method and risk components of the luggage information asset.

The motivation of the luggage check-in personnel to carry out the first attack is to sabotage the safety of the flight. The attacker has knowledge of how the luggage check-in process works, and also has access to the documents for recording details about the luggage. To carry out the attack, the personnel have to manipulate the records in the luggage document so that the aircraft will be overloaded.

The motivation of the personnel to carry out the second attack on the airline is to hide contraband item in the passenger luggage. This he can achieve because he has physical access to the passenger luggage and also knows how the luggage check-in process works.

The vulnerabilities present in the luggage check-in process for these two attacks to be possible is as follows - data generated by the check-in personnel is not verified and activities of the check-in personnel are not properly monitored.

The impact of these two attacks are as follows - for the first attack, its leads to loss of integrity in data contained in luggage information and also leads to overloading the aircraft. For the second attack, the impact is that passengers loose trust in the luggage check-in process and loss of integrity of the contents of passengers' luggage.

Table 6. Luggage information risk- and threat analysis.

	Risk 3	Risk 4
Threat agent	Luggage check-in personnel Motivation: Sabotage the safety of the flight Resources: weighing machine, luggage information document Expertise: Knowledge about luggage check-in process	Luggage check-in personnel Motivation: hide contraband item on passenger luggage Resources: has physical access to the luggage Expertise: knowledge of the check-in process and messaging system
Attack method	Personnel accept luggage and measure luggage Records values lower than actual weight of luggage Sends luggage information to ground services for onward loading of the aircraft	Personnel accept and weigh luggage Adds contraband item to passenger luggage and record the weight of luggage Sends luggage information to ground services for onward loading of the aircraft
Threat	Personnel measures and records values lower than actual weight of luggage, sends luggage and luggage information to ground operations for onward loading of the aircraft	Personnel accepts luggage and adds contraband item to passengers luggage, sends luggage and luggage information to ground operations to load the aircraft
Vulnerability	Luggage information generated by the check-in personnel is not verified	Personnel activity is not monitored
Event	Personnel records values lower than actual weight of luggage, and ground operations uses the information in the loading of the aircraft because luggage information generated by personnel is not verified	Personnel accepts luggage and adds contraband item to passengers luggage, sends luggage and luggage information to ground operations to load the aircraft because personnel activity is not monitored
Impact	Loss of integrity of luggage information Incorrect calculation of centre of gravity (cg) of aircraft	Loss of integrity of contents of passenger luggage Loss of trust in luggage check-in process
Risk	Personnel records values lower than actual weight of luggage, and ground operations uses the information in the loading of the aircraft because luggage information which results to loss of integrity of luggage information and incorrect calculation of cg of which might cause aircraft to be unstable	Personnel accepts luggage and adds contraband item to passengers luggage, sends luggage and luggage information to ground operations to load the aircraft because personnel activity is not monitored which results to loss of integrity of contents of passenger luggage and loss of trust in luggage check-in process

4.3   Analysis of Risks to the *Fuelslip* Asset

For the fuel-slip risk analysis, two possible attack methods are described for this business asset. For each of these attacks, two possible threat agents are identified -

a malicious insider and an arbitrary attacker. The Table 7 below describes the threat agents, the attack methods and risk components of the fuelslip asset.

From the Table 7, the goal of the malicious insider to carry out the first attack is to sabotage the safety of the flight. The Malicious insider has access

Table 7. Fuelslip risk- and threat analysis.

	Risk 5	Risk 6
Threat agent	Malicious insider Motivation: sabotage the safety of the flight Resources: access to fuelslip document Expertise: knowledge of refueling process	An attacker Motivation: sabotage the safety of the flight Resources: access to airlines messaging system and mailing list Expertise: knowledge of refueling process
Attack method	Malicious insider accesses computer storing fuelslip documents Makes changes to the content of fuelslip Fuelslip is sent to the service provider	Attacker intercepts fuelslip Airline sends fuelslip to attacker Attacker changes data contained in fuelslip Attacker sends edited fuelslip to supplier Supplier conducts refueling based on information in fuelslip received
Threat	Malicious insider access the fuelslip document and changes the data contained in the document	Attacker intercepts fuelslip, receives fuelslip, changes data contained, sends to supplier, refueling is conducted based on information on fuelslip
Vulnerability	Fuelslip document is not encrypted	Email message can be intercepted
Event	Malicious insider with access to computer that stores fuelslip make changes to the data contained in fuelslip before it is sent to service provider because the document is not encrypted	Attacker intercepts fuelslip, receives fuelslip, changes data contained, sends to supplier, refueling is conducted based on information on fuelslip because email message can be intercepted
Impact	Loss of integrity of fuelslip Lower quantity or different type of fuel can be loaded to the aircraft	Loss of integrity of fuelslip Data contained in fuelslip can be changed
Risk	Malicious insider with access to computer that stores fuelslip make changes to the data contained in fuelslip before it is sent to service provider because the document is not encrypted which results to loss of integrity of fuelslip and can cause the aircraft to be loaded with wrong quantity and type of fuel	Attacker intercepts fuelslip, changes data contained, sends to supplier, refueling is conducted based on information on fuelslip because messaging system can spoofed, which causes loss of integrity of fuelslip and can result in loading the aircraft with wrong quantity and type of fuel

to fuelslip document and has expert knowledge of the airline refueling process. The malicious insider can make changes to the fuelslip document, and when the service provider loads the refuels the aircraft with the service requirement in the fuel, the aircraft will be loaded with the wrong fuel.

For the second attack, a random attacker with a help of an insider can manipulate the messaging system and intercept fuelslip document. The motivation is the same and that is to sabotage the safety of the aircraft. The fuelslip is transmitted via the messaging system to the service provider. An attacker (with the help of an insider), can intercept the fuelslip and modify data contained in it. In the event of successful exploitation of this weakness, the effect is that the aircraft is loaded with the wrong type or quantity of fuel.

The weakness in the system for these attacks to be possible is because the fuelslip document is not encrypted and the email messages between airline and service providers can be intercepted. These two attacks negatively affect the airline by causing loss of integrity of fuelslip document and also possibly resulting in the aircraft to be loaded with the wrong fuel.

4.4   Analysis of Risks to the *Cargo Assignment* Asset

There are two possible attacks that are described for cargo assignment business asset. The threat agents for each of these attacks are malicious insider and a random attacker. The Table 8 below describes each of the threat agent, the attack method and risk components of the cargo assignment asset.

The goals of the two attacks described in the Table 8 above are the same, to sabotage the safety of the aircraft. For the first attack, the malicious insider has access to the cargo assignment document and also has the knowledge of aircraft loading process. To carry out the attack, the malicious insider changes data contained in fuelslip before it is sent to the service provider to reload the aircraft. The service provider loads the aircraft with the information in the cargo assignment.

The second attack involves intercepting of the cargo assignment document as it sent from airline to the service provider. This the attacker can achieve by hacking the airline mailing list so that he can receive email sent to the service providers. By intercepting and changing data contained in cargo assignment before it gets to the service provider, the aircraft will be reloaded improperly.

The vulnerabilities that results in these attacks are as follows - the cargo assignment document is not encrypted and the mailing list is not properly secured. These attacks will affect the airline negatively because of the loss of integrity of fuelslip document resulting from the attack. Also, the attack can cause the aircraft to be improperly loaded.

Table 8. Cargo assignment risk- and threat analysis.

	Risk 7	Risk 8
Threat agent	Malicious insider Motivation: sabotage the safety of the flight Resources: access to cargo assignment document Expertise: knowledge of luggage and cargo loading process	An attacker Motivation: sabotage the safety of the flight Resources: access to airlines messaging system and mailing list Expertise: knowledge of cargo loading system
Attack method	Access the cargo assignment document Make changes to the cargo assignment document Changed cargo assignment is sent to service provider	Attacker hacks airlines mailing list Attacker replaces service provider email with his email Airline sends cargo assignment to attacker Attacker changes data contained in cargo assignment Attacker sends edited assignment to service provider Service provider conducts cargo and luggage loading based on information in cargo assignment document received
Threat	Malicious insider with access to cargo assignment document make changes to cargo assignment document before it is sent to service provider	Attacker hacks airline mailing list, receives cargo assignment, changes data contained, sends to service provider, loading is conducted based on information on cargo assignment
Vulnerability	Cargo assignment document is not encrypted	Mailing list is not fully secured
Event	Malicious insider with access to cargo assignment document make changes to cargo assignment document before it is sent to service provider because cargo assignment document is not encrypted	Attacker hacks airline mailing list, receives cargo assignment, changes data contained, sends to service provider, loading is conducted based on information on cargo assignment because mailing list is not fully secured
Impact	Loss of integrity of cargo assignment document Aircraft is not properly loaded	Loss of cargo assignment slip Data contained in cargo assignment can be changed
Risk	Malicious insider with access to cargo assignment document make changes to cargo assignment document before it is sent to service provider because cargo assignment document is not encrypted which causes loss of integrity of cargo assignment and improper loading of the aircraft and can result to instability of the aircraft in the air	Attacker hacks airline mailing list, receives cargo assignment, changes data contained, sends to service provider, loading is conducted based on information in cargo assignment, because mailing list is not fully secured which causes loss of integrity of cargo assignment and can result to overloading the aircraft

5   Security Risk Mitigation

In this section, firstly, we cover the three last steps presented in Fig. 3. In all cases, to mitigate the identified security risks, the *risk reduction* treatment decision is taken. The latter we refine to security requirements for implementation with existing security controls.

Security requirements and controls suggestions to mitigate the identified risks we provide in Table 9. The security requirement needed to reduce *Risk 1* is monitoring the activities of check-in personnel. In order to achieve this security requirement, a control is applied. The security control requires an additional

Table 9. Security requirements and counter-measures.

Risk	Security requirement	Control description
Risk 1	Monitor the activity of check-in personnel	Officer verifying the actions of check-in personnel
Risk 2	Educate the airline passengers on phishing attacks	Using secured https websites for booking and check-in activity
Risk 3	Monitor activities of luggage check-in personnel	Random checks to verify weight records with actual weight of luggage
Risk 4	Monitor activities of luggage check-in personnel	Install camera to record luggage check-in personnel activities
Risk 5	Access control on fuel slip document	Encrypt fuel slip document
Risk 6	Make information contained in fuel slip unreadable	Verifies received document with previous originals received (Blockchain cryptographic digest PKI or PGP) Encrypt fuel slip document
Risk 7	Access control on cargo assignment document	Encrypt cargo assignment document
Risk 8	Make information contained in cargo assignment unreadable	Verify received document with previous originals received. Encrypt cargo assignment document

officer to always verify the activities for the check-in personnel. The cost value of 4 implies that it costs more to employ additional staff to verify the activities of check-in personnel.

For *Risk 2*, the security requirement to reduce the risk is achieved by educating airline passengers on possible phishing-attack methods. To achieve this, a security control is applied by using only secured https websites for booking and passenger check-in. The cost value of 2 implies that it costs less to educate the passengers against phishing methods and provide secured website for booking and check-in activities.

The security requirement for reducing *Risk 3* is monitoring activities of luggage check-in personnel. The security control that must be implemented to achieve this is performing random checks on activities of luggage check-in personnel. The risk treatment value of 2 implies that less cost is required to perform checks that verify data recorded by personnel.

For *Risk 4*, the security requirement is the same as for Risk 3 which is monitoring activities of personnel. However, a different security control is applied. The control is achieved by installing cameras to record activities of luggage check-in personnel and scanning the recordings. The treatment cost of 1 shows that little money is required to mount security cameras to monitor activities of check-in personnel.

The security requirement for *Risk 5* is controlling access to the fuel slip document. Access control for the fuel slip is achieved by encrypting the fuel slip document. Such encryption that relies on a public-key infrastructure (PKI) is applicable in this case. The fuel slips and other documents that contain service requirements are encrypted with private keys of the selected supplier and therefore, only the supplier views the document, even when intercepted by another person.

For *Risk 6*, the same security requirement and security control apply as in *Risk 5*. The risk treatment value of 4 for both risks implies that it costs the airline a lot to implement an encryption that uses PKI.

The security requirements and controls for *Risk 7* and *Risk 8* are the same. In order to reduce these risks for the cargo-assignment document, proper access control must be implemented. The latter is achieved by applying encryption that relies on PKI. As a result, only the service provider can have access to the document, even when interception occurs. The cost value of 4 implies that it is expensive to implement encryption that depends on PKI in order to reduce *Risk 7* and *Risk 8*.

6   Risk Assessment and Control Selection

In this section, we return to Step 3 with a focus on *security risk assessment* and Step 6 with a focus on *control selection* that are presented in Fig. 3. Hence we assess how the security requirements and controls of Sect. 5 affect the identified security risks of Sect. 4. For that assessment, we use the goal question metric (GQM) [26]. The ISSRM approach [17] suggests a number of questions to maximize risk reduction and minimize risk-treatment costs. In this context, our emphasis is placed on risk-reduction estimation.

6.1   Security Risk Assessment

The GQM-questions target the risk level, its occurrence frequency, importance regarding the business, the risk-reduction level after treatment of risk. The risk management metrics, such as business asset value, threat likelihood, vulnerability level, and security objective are estimated in the scale from 0 (lowest) to 5 (highest). The risk event, risk impact and risk level are then calculated as follows:

- Risk event = threat likelihood + vulnerability level − 1
- Impact = maximum value of the security criterion
- Risk level = risk event x impact.
- Maximum-risk level = $(5 + 5 − 1) * 5 = 45$
- Minimum-risk level = $(0 + 0 − 1) * 0 = 0$
- Risk reduction level = Risk level 1 − Risk level 2

Table 10. Risk metrics before and after risk treatment.

	Before treatment					After treatment				Risk reduction level	Business asset value	Cost of counter-measure
	Vulnerability level	Threat likelihood	Event potentiality	Impact level	Risk level1	Vulnerability level	Threat likelihood	Event potentiality	Risk level2			
Risk1	3	2	4	3	12	2	1	2	6	6	3	4
Risk2	2	4	5	3	15	1	3	3	9	6	3	2
Risk3	1	2	2	2	4	1	1	1	2	2	1	2
Risk4	4	2	5	3	15	2	1	2	6	9	1	1
Risk5	3	3	5	4	20	1	1	1	4	16	3	4
Risk6	3	2	4	4	16	1	1	1	4	12	3	4
Risk7	2	3	4	3	12	1	1	1	4	8	1	4
Risk8	2	2	3	3	9	1	2	2	6	3	1	4

The minimum-risk level obtainable is 0, while the maximum-risk level obtainable is 45. Therefore, 0 and 45 represent the boundaries of the risks. Here, Risk level 1 is calculated when no security countermeasures are applied and Risk level 2 is calculated after the application of the security countermeasures. The collected data are illustrated in Table 10.

6.2   Security Control Selection

Not all security risks of our study can be mitigated, e.g., because of lacking resources, or time-to-market requirements. Thus, security control selection, potentially, means understanding which controls need to be selected, or in other words which security risks need to be mitigated first. One way to perform this trade-off analysis is by using the *value* of the business asset, *counter-measure cost* and *risk reduction level* (RRL), gathered in Table 10 (see, three last columns). Using these metrics, three graphs are prepared (see Figs. 4, 5 and 6) including data on RRL and value, RRL and cost, and cost and value. The graphs are divided into four quadrants and the priority on each quadrant is identified by labels low (L), medium (M) and high (H) on each quadrant.

Figure 4 shows a graph about the risk-reduction level against business asset value. The desired situation is a high value asset with a high risk reduction value. This can be identified in the quadrant that has Risk 6 (R6), R5 and therefore represents a high priority. The medium priorities quadrants have high

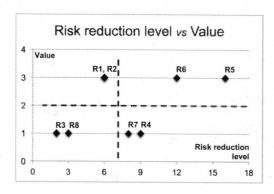

Fig. 4. Risk-reduction level against business asset value.

asset value with low risk-reduction level and low-value assets coupled with a high risk-reduction values. These situations are found in quadrants that have R1, R2 and R7, R4 respectively. The least desired situation is a low valued asset with a low risk reduction in quadrant that has R3 and R8.

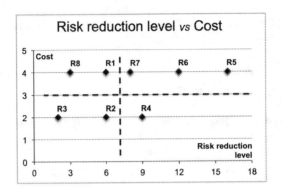

Fig. 5. Risk-reduction level against cost of counter measure.

Figure 5 shows a graph of risk-reduction level against cost of counter measure. The ideal situation is a low cost value with a high risk reduction value. This can be identified in the quadrant comprising R4 and therefore, it represents a high priority. The medium-priority quadrants have high cost value with high risk-reduction level and low cost with low risk-reduction values. These situations are found in quadrants that comprise R5, R6, R7 and R2, R3 respectively. The low priority can be identified in the quadrant of high cost and low risk reduction. This quadrant contains R1 and R8.

Figure 6 is about the cost of counter measure against business asset value. A low cost treatment with a high-value asset represents a high priority and can be seen in the quadrant with R2. The medium priority are found in quadrants combining high-value assets with high cost of counter measure and low-value assets and low cost of counter measure. These are found in the quadrant comprising R1, R5, R6 and the quadrant with R3, R4. The least ideal situation is a low-value asset with a high cost of risk treatment and it is found in the quadrant with R7, R8.

Table 11 shows risk priorities derived from combining the graphs of Figs. 4, 5 and 6 where a value of 1 is assigned to low priority risks, medium priority risks has a value of 2, while the value of 3 is assigned to high priority risks. By adding these values across the three graphs, a priority can be estimated that depends on the value of a business asset, cost of counter measure and risk reduction level. The risks with high priorities are R2, R4, R5 and R6. The medium priority risks are R1, R3 and R7. The least priority risk is R8.

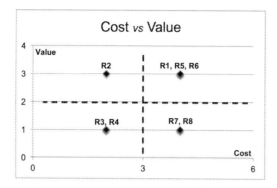

Fig. 6. Cost of counter measure against business asset value.

Table 11. Risk versus priority.

	Value-RRL	RRL-cost	Value-cost		
	Graph 1	Graph 2	Graph 3		
Risk 1	2	1	2	5	Medium priority
Risk 2	2	2	3	7	High priority
Risk 3	1	2	2	5	Medium priority
Risk 4	2	3	2	7	High priority
Risk 5	3	2	2	7	High priority
Risk 6	3	2	2	7	High priority
Risk 7	2	2	1	5	Medium priority
Risk 8	1	1	1	3	Low priority

7   Risk Simulation and Analysis of Simulation Results

We simulate one of the risks described in Sect. 4 where the simulated risk is chosen based on the risk priority identified in Table 8. The risk with the highest priority is chosen for simulation. The goal of the simulation is to demonstrate how an attack on airline turnaround operations can be carried out by exploiting one of the identified risks. The simulation also shows the success of the attack before and after the application of security controls. The effects of the attack on airline resources are also observed.

The airline resources - fuelslip, aircraft and pilot are taken as inputs for simulation. The inputs passes through the stages of the airline refueling process. These are repeated over a certain number of iteration and the effects of the attack are observed. The outputs of the simulation are represented by the number of unsuccessful attacks and successful attacks (emergency landing and aircrash). The remainder of this section is structured as follows. First, Sect. 7.1 describes the encountered risks, followed by Sect. 7.2 where we introduce the simulation platform.

7.1   Description of Risk

Risk 4 is chosen for the simulation and it involves an exchange of service requirements between the airline and service provider using a messaging system. This
risk can easily be represented in a computer based environment for simulation
purpose.

Risk 4 *An attacker intercepts fuelslip, changes data contained, sends to supplier, refueling is conducted based on information on fuelslip because messaging
system can spoofed, which causes loss of integrity of fuelslip and can result in
loading the aircraft with wrong quantity and type of fuel.*

One of the methods that can be used to intercept information exchanged
between two parties is DNS poison and we briefly describe this method. A malicious insider that has access to the DNS server can make changes to information
contained in the DNS system and thereby divert an email to a malicious attacker.
It is possible that an insider changes the IP address of the supplier domain to
that of the attacker. When an email is sent to the supplier from the airline,
the SMTP server delivers the email to the attacker email server. The attacker
receives the email, make changes to the service requirements in the fuelslip and
forward to the service provider.

7.2   Simulation Platform

The simulation is carried out using Anylogic simulation toolkit[3] (student edition). Anylogic provides process modeling blocks for modeling and describing
discrete events. The Process Modeling in anylogic is a collection of objects for
defining process workflows and their associated resources.

Cases for Simulation: The attack is described above simulated in two scenarios. The first case shows how the attack is achieved and the result of the attack
before security control is applied on the risk in Fig. 7. The second case shows the
attack and the corresponding results after security control is applied in Fig. 8.

Before Application of Security Control: Table 12 describes activities that
are generated for the airline refueling process before the application of security
controls. A role is attached for each of the activities, and also various input(s)
and outputs(s) are listed for the activity.

After Application of Security Control: Table 13 describes activities that are
generated for airline refueling process after the application of security controls. A
role is attached for each of the activities, and also various input(s) and outputs(s)
are listed for the activities.

---

[3] http://www.anylogic.com/.

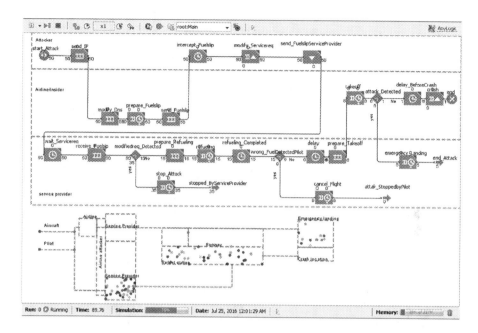

Fig. 7. Airline-refuel attack before the application of security controls.

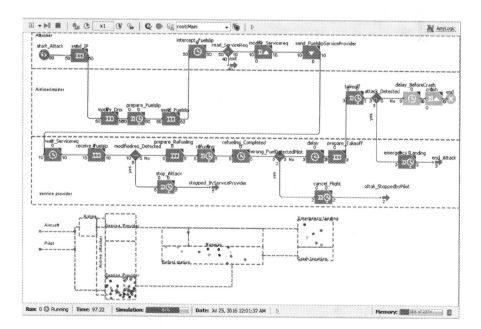

Fig. 8. Airline-refuel attack after the application of security controls.

Table 12. Simulation Process Description before security control application

Activity	Description	Role	Input	Output
StartAttack	Attacker Starts the attack process	Attacker	–	Pilot, Aircraft, Fuelslip
SendIP	Attacker sends IP address to the insider agent	Attacker	Pilot, Aircraft, Fuelslip	IpAddress
ModifyDns	Insider agent modifies DNS with attacker IP details	Insider	IpAddress	IpAddress
PrepareFuelslip	Airline prepares fuelslip	Insider	IpAddress	IpAddress
SendFuelslip	Airline sends fuelslip to service provider	Airline	Fuelslip	Fuelslip
InterceptFuelslip	Attacker intercepts fuelslip	Attacker	Fuelslip	Fuelslip
ModifyServicereq	Attacker modifies service requirement in fuelslip	Attacker	Fuelslip	Fuelslip
SendFuelService Provider	Attacker sends modified fuelslip to service provider	Attacker	Fuelslip	Fuelslip
WaitServicereq	Service provider waits for refueling service requirement	Service provider	Fuelslip	Fuelslip
ReceiveFuelslip	Service provider receives fuelslip	Service provider	Fuelslip	Fuelslip
Modifyreq–Detected	Chances that service provider detects changes in fuelslip	Service provider	Fuelslip	Fuelslip
StopAttack	Service provider prepares to stop attack	Service provider	Fuelslip	Fuelslip
StoppedByService Provider	Process exits as attacked is stopped by Service provider	Service provider	Fuelslip	–
PrepareRefueling	Service provider prepares aircraft for refueling	Service provider	Fuelslip, Aircraft	Aircraft
Refueling	Refueling process by service provider	Service provider	Aircraft	Aircraft
Refueling–Completed	Refueling is completed	Service provider	Aircraft	Aircraft
WrongFuel–DetectedPilot	Pilot detects wrong fuel in the aircraft	Airline	Aircraft, Pilot	Aircraft, Pilot
CancelFlight	Pilot cancels flight	Airline	Aircraft, Pilot	Aircraft, Pilot
AttackStoppedby–Pilot	Attack is stopped by pilot	Airline	Aircraft	–
Delay	Delay before takeoff	Airline	Aircraft, Pilot	Aircraft, Pilot
PrepareTakeoff	Pilot prepares for takeoff	Airline	Aircraft, Pilot	Aircraft, Pilot
Takeoff	Aircraft takeoff	Airline	Aircraft, Pilot	Aircraft, Pilot
AttackDetected	Probability that pilot detects abnormality on the aircraft	Airline	Aircraft, Pilot	Aircraft, Pilot
Emergency–Landing	Emergency landing due to detected abnormality	Airline	Aircraft, Pilot	Aircraft, Pilot
EndAttack	Attack ends	Airline	Aircraft	–
DelayBeforeCrash	Aircraft flies with wrong fuel in the tank	Airline	Aircraft, Pilot	Aircraft, Pilot
Crash	Flying with wrong fuel results in aircrash	Airline	Aircraft	–

Table 13. Simulation Process Description after security control application

Activity	Description	Role	Input	Output
StartAttack	Attacker starts the attack process	Attacker	–	Pilot, Aircraft, Fuelslip
SendIP	Attacker sends Ip address to an insider agent	Attacker	Pilot, Aircraft, Fuelslip	IpAddress
ModifyDns	Insider agent modifies DNS with attacker IP details	Insider	IpAddress	IpAddress
PrepareFuelslip	Airline prepares fuelslip	Insider	IpAddress	IpAddress
SendFuelslip	Airline sends fuelslip to service provider	Airline	Fuelslip	Fuelslip
ReadService	Probability that the attacker can decrypt encrypted fuelslip	Attacker	Fuelslip	Fuelslip
Exit	Attack ends as attacker can't view fuelslip content	Attacker	Fuelslip	–
InterceptFuelslip	Attacker intercepts fuelslip	Attacker	Fuelslip	Fuelslip
ModifyServicereq	Attacker modifies service requirement in fuelslip	Attacker	Fuelslip	Fuelslip
SendFuelService Provider	Attacker sends modified fuelslip to service provider	Attacker	Fuelslip	Fuelslip
WaitServicereq	Service provider waits for refueling service requirement	Service provider	Fuelslip	Fuelslip
ReceiveFuelslip	Service provider receives fuelslip	Service provider	Fuelslip	Fuelslip
ModifyreqDetected	Chances that service provider detects changes in fuelslip	Service provider	Fuelslip	Fuelslip
StopAttack	Service provider prepares to stop attack	Service provider	Fuelslip	Fuelslip
StoppedByService Provider	Process exits as attacked is stopped by Service provider	Service provider	Fuelslip	–
PrepareRefueling	Service prepares aircraft for refueling	Service provider	Fuelslip, Aircraft	Aircraft
Refueling	Refueling process by service provider	Service provider	Aircraft	Aircraft
Refueling–Completed	Refueling is completed	Service provider	Aircraft	Aircraft
WrongFuel–Detected Pilot	Pilot detects wrong fuel in the aircraft	Airline	Aircraft, Pilot	Aircraft, Pilot
CancelFlight	Pilot cancels flight	Airline	Aircraft, Pilot	Aircraft, Pilot
AttackStoppedby–Pilot	Attack is stopped by pilot	Airline	Aircraft	–
Delay	delay before takeoff	Airline	Aircraft, Pilot	Aircraft, Pilot
PrepareTakeoff	Pilot prepares for takeoff	Airline	Aircraft, Pilot	Aircraft, Pilot
Takeoff	Aircraft takeoff	Airline	Aircraft, Pilot	Aircraft, Pilot
AttackDetected	Probability that pilot detects abnormality on the aircraft	Airline	Aircraft, Pilot	Aircraft, Pilot
Emergency–Landing	Emergency landing due to detected abnormality	Airline	Aircraft, Pilot	Aircraft, Pilot
EndAttack	Attack ends	Airline	Aircraft	–
DelayBeforeCrash	Aircraft flies with wrong fuel in the tank	Airline	Aircraft, Pilot	Aircraft, Pilot
Crash	Aircraft crashes and attack comes to an end	Airline	Aircraft	–

7.3   Assessment of Simulation

Table 14 shows the end result of the attack after a specific number of attempts in percentage. Before the security control was applied. 70% are stopped by service provider after detecting changes in fuelslip. The pilot stops 18% of the attack after detecting wrong fuel during the refueling process. About 12% of the attack is capable of causing physical damage to aircraft and possibly loss of life of the passengers; where 10% resulted in emergency landing and 2% leads to air crash.

Table 14. Airline refueling simulation result.

Airline refuel attack	Before security control	Percentage	After security control	Percentage
Total number of attempts	50	100	50	100
Attack stopped by encryption	–	–	40	80
Attack stopped by service provider	35	70	5	10
Attack stopped by pilot	9	18	2	4
Attack ends in emergency landing	5	10	3	6
Attack ends in aircrash	1	2	0	0

After the application of security controls, 80% of the attack is stopped due to encryption of fuelslip. The service provider stops about 10% of the attack and 4% is stopped by the pilot. Only 6% of the attack cause physical damage on the aircraft as a result of emergency landing.

Table 15 shows airline resources that are affected by various endings of the attack simulation. If the attack is stopped the service provider, the only resources affected is the fuelslip. This could have little economic damage on the airline as a result of delay experienced by the airline in using alternative means to deliver the refueling service requirement to service provider.

Table 15. Effect of attack on airline resources.

Resources	Stopped by service provider	Stopped by pilot	Emergency landing	Aircrash	Stopped by encryption
Fuelslip	*	*	*	*	*
Pilot	–	*	*	*	–
Aircraft	–	*	*	*	–
Runway	–	–	*	*	–
Control tower	–	–	*	*	–
Fire service	–	–	*	*	–

If the attack is stopped by pilot, the following resources fuelslip, pilot, and aircraft are affected by the attack. This could cause a significant economic loss resulting from the waiting time in emptying and refueling the aircraft with proper fuel. In addition, economic loss in finding alternative means of sending service requirements since the messaging system is compromised.

An attack ending in emergency landing may cause physical damage to the aircraft, pilot and passengers as well. Other resources belonging to the airport such as the runway, fire service and control tower are also affected by the attack. This results in a considerable economic loss, not only to the airline but also to the airport as well. An attack ending aircrash is the worst situation envisaged that results in the total destruction of the aircraft, loss of human life and considerable loss to the airline.

When a security control is applied and the attack is ended as a result of encryption, no airline resources are affected and no economic loss results from the attack. This is best situation anticipated, the refueling process executes perfectly with no hitches.

Therefore, if the fuelslip is not encrypted and the attack starts to execute, there must be a negative consequence on the airline irrespective of how the attack ends. This is as a result of seized airline (and airport) resources, delays experienced, damages on the aircraft and possibly loss of human lives.

7.4   Related Studies on Risk Assessment in Aviation Sector

In [13] a quantitative analysis method is applied to evaluate risks associated to civil aviation sector. The effect of factors such as human error, mechanical failures, terrorist actions, crew experience in causing an accident was examined. The result of the analysis showed the rate of fatal accidents that can occur overtime in civil aviation sector.

In [9] authors propose a combination of qualitative and quantitative approaches to tackle security in civil aviation. This was achieved by applying Fault Tree (FT), risk assessment and method statements (RAMS) analysis methods. The areas considered in the analysis include access control on airport structures, baggage handling processes, aircraft and flight procedures, cargo and catering processes. RAMS assessment method was applied to determine the threat level, vulnerability level and amount of risks. The FT method was applied to examine possible attack methods and evaluate their impacts. The result is an optimization of protection levels for airport security.

There are limitations both in [9,13]. The processes and activities involved in the aviation sector are numerous and with various complexities [9,13]. In our work, we concentrate on risks in a normalized turnaround processes of the aviation sector. The processes include gate operations, passenger management and ground operations. While the result of risk assessment carried out in [13] shows the amount of possible accidents that can occur over a period of time in the sector, it does not provide any means for combating and treating the issues identified as possible cause for these accidents. Also, in [9], threat assessment were performed on possible attack scenario to determine attack impacts, the work does not show how these threats can be eliminated or reduced. In our study, risks were identified on assets in airline turnaround processes; security requirements, security controls are applied and evaluated to determine the risk reduction levels.

The result of [13], [9] and our study show some similarities. The result of [13] shows the rate of fatal air crashes leading to loss of life. Our simulation result shows the number of attack that can lead to plane crashes and emergency landing over a certain amount of repetition. While [13] considers outcome only from rate of fatal accident resulting from an existing risk factors, our study provides two level results. One from existing risk factors and another after security control has been applied.

Some results received in [9] correspond to the passenger management and ground operation pools in airline turnaround operations in our study. While different approaches for risk assessment were used in [9] and in our work, in both studies the threat level, vulnerabilities and amount of risks were evaluated. The outcome of [9] have optimized protection levels for airport security. This is comparable to the risk reduction obtained after security control was applied in the turnaround processes. This is evident by comparing results of the simulations before and after security control has been applied. The result of the simulations show a decrease in successful attacks on the asset. The risk reduction corresponds to the optimized security level of the process simulated.

8   Conclusions

In this paper, security issues affecting cross-organizational collaborations are analyzed using the aviation sector. The airline turnaround process presents a suitable environment for this study as operations are resource intensive. The end result of the analysis in this work is a set of security requirement- and controls for managing risks resulting from the collaboration between airlines and service providers.

The relevant assets for collaborations between enterprises are identified by describing airline turnaround processes. We identify the IS assets in the passenger-management process and ground operations. The risks identified in the risk analysis section are reduced by applying security requirement and controls for each risk identified. We find the security requirements for the reduction of risks identified for the checked-in passenger information, the luggage information, the security requirements. Likewise, we detect the security requirements for the reduction of risks identified in the ground operations for the fuel slip and the cargo assignment.

The security controls for the reduction of risks are listed as follows: For the checked-in passenger information, the security controls are an officer verifying the actions of check-in personnel and using only secured https websites for booking and check-in activity. For the luggage information, the security controls are random checks to verify weight records with actual weight of luggage and installing cameras to record luggage check-in personnel activities. The security controls for the reduction of risks identified in ground operations are listed as follows. For the fuel slip, the security controls are encrypting fuel-slip documents, and verifying all received documents and comparing them with previously received originals. For the cargo assignment, the security controls are encrypt the cargo-assignment document, and verify all received documents and compare with previous originals received. Furthermore, we show the degree of security that is achieved with implementing security controls.

There are few limitations observed in this work. The risk analysis carried out is limited by the scope of data collection. Interviews, discussions and surveys were carried out with experts in the aviation industry. However, the number of individuals involved is not complete enough and requires follow-up studies. Also, by using a normalized airline-turnaround processes as a case study for the analysis carried out, this work should be expanded also to additional support processes. The airline turnaround operations are very complex and involve more activities that are not represented in the normalized processes. Therefore this work does not represent the full reality of risk assessment in airline turnaround operations. Finally, this work does not produce a secure cross-organizational business process as one of the initial goals this work was set to achieve.

As future research we plan to apply risk-based patterns in modeling a secure business process for cross-organizational collaborations and also intend to analyze security threats in cloud-supported enterprise collaborations. Additionally, we intend to further explore methods for improved risk-metrics calculations in a cross–organizational business process that will properly examine the impact of risks on different collaborating parties.

References

1. US Department of Transportation: Aircraft weight and balance handbook (2007). http://tiny.cc/m7xkcy
2. NATA Safety 1st eToolkit (2015). http://tiny.cc/5nzkcy
3. Anton, V.U., Eduardo, B.F.: An extensible pattern-based library and taxonomy of security threats for distributed systems. In: Security in Information Systems: Advances and New Challenges, vol. 36, pp. 734–747 (2014)
4. Bartelt, C., Rausch, A., Rehfeldt, K.: Quo vadis cyber-physical systems: research areas of cyber-physical ecosystems: a position paper. In: Proceedings of the 1st International Workshop on Control Theory for Software Engineering, CTSE 2015, pp. 22–25. ACM, New York (2015)
5. Belobaba, P., Odoni, A., Barnhart, C.: The Global Airline Industry. Wiley, Chichester (2015)
6. Dirk, D.: Smart business process management. In: Workflow Management Coalition, pp. 207–223 (2012)
7. Draheim, D.: Business Process Technology: A Unified View on Business Processes, Workflows and Enterprise Applications. Springer, Heidelberg (2010). https://doi.org/10.1007/978-3-642-01588-5

8. Dubois, É., Heymans, P., Mayer, N., Matulevičius, R.: A systematic approach to define the domain of information system security risk management. In: Nurcan, S., Salinesi, C., Souveyet, C., Ralyté, J. (eds.) Intentional Perspectives on Information Systems Engineering, pp. 289–306. Springer, Heidelberg (2010). https://doi.org/10.1007/978-3-642-12544-7_16

9. Tamasi, G., Demichela, M.: Risk assessment techniques for civil aviation security. J. Reliab. Eng. Syst. Saf. **96**, 892–899 (2011)

10. Kutvonen, L., Norta, A., Ruohomaa, S.: Inter-enterprise business transaction management in open service ecosystems. In: 2012 IEEE 16th International Enterprise Distributed Object Computing Conference (EDOC), pp. 31–40. IEEE (2012)

11. Leonardi, M., Piracci, E., Galati, G.: Ads-b vulnerability to low cost jammers: risk assessment and possible solutions. In: 2014 Tyrrhenian International Workshop on Digital Communications-Enhanced Surveillance of Aircraft and Vehicles (TIWDC/ESAV), pp. 41–46. IEEE (2014)

12. Long, S.: Socioanalytic Methods: Discovering the Hidden in Organisations and Social Systems. Karnac Books, London (2013)

13. Janic, M.: An assessment of risk and safety in civil aviation. J. Air Transp. Manag. **6**, 43–50 (2000)

14. Maiden, N., Ncube, C., Lockerbie, J.: Inventing requirements: experiences with an airport operations system. In: Paech, B., Rolland, C. (eds.) REFSQ 2008. LNCS, vol. 5025, pp. 58–72. Springer, Heidelberg (2008). https://doi.org/10.1007/978-3-540-69062-7_6

15. Massacci, F., Paci, F., Tedeschi, A.: Assessing a requirements evolution approach: empirical studies in the air traffic management domain. J. Syst. Softw. **95**, 70–88 (2014)

16. Matulevičius, R., Norta, A., Udokwu, C., Nõukas, R.: Security risk management in the aviation turnaround sector. In: Dang, T.K., Wagner, R., Küng, J., Thoai, N., Takizawa, M., Neuhold, E. (eds.) FDSE 2016. LNCS, vol. 10018, pp. 119–140. Springer, Cham (2016). https://doi.org/10.1007/978-3-319-48057-2_8

17. Mayer, N.: Model-based management of information system security risk. Ph.D. thesis, University of Namur (2009)

18. Business Process Model. Notation (bpmn) version 2.0. Object Management Group specification (2011). http://www.bpmn.org

19. Nõukas, R.: Service brokering environment for an airline. Master thesis, Tallinn University of Technology (2015)

20. Norta, A.: Creation of smart-contracting collaborations for decentralized autonomous organizations. In: Matulevičius, R., Dumas, M. (eds.) BIR 2015. LNBIP, vol. 229, pp. 3–17. Springer, Cham (2015). https://doi.org/10.1007/978-3-319-21915-8_1

21. Norta, A., Grefen, P., Narendra, N.C.: A reference architecture for managing dynamic inter-organizational business processes. Data Knowl. Eng. **91**, 52–89 (2014)

22. Norta, A., Ma, L., Duan, Y., Rull, A., Kõlvart, M., Taveter, K.: eContractual choreography-language properties towards cross-organizational business collaboration. J. Internet Serv. Appl. **6**(1), 1–23 (2015)

23. Samarütel, S., Matulevičius, R., Norta, A., Nõukas, R.: Securing airline-turnaround processes using security risk-oriented patterns. In: Horkoff, J., Jeusfeld, M.A., Persson, A. (eds.) PoEM 2016. LNBIP, vol. 267, pp. 209–224. Springer, Cham (2016). https://doi.org/10.1007/978-3-319-48393-1_15

24. Sampigethaya, K., Poovendran, R.: Aviation cyber-physical systems: foundations for future aircraft and air transport. Proc. IEEE **101**(8), 1834–1855 (2013)

25. Shim, W., Massacci, F., Tedeschi, A., Pollini, A.: A relative cost-benefit approach for evaluating alternative airport security policies. In: 2014 Ninth International Conference on Availability, Reliability and Security (ARES), pp. 514–522. IEEE (2014)
26. van Solingen, R., Basili, V., Caldiera, G., Rombach, H.D.: Goal Question Metric (GQM) Approach. Wiley, Hoboken (2002)

Scalable Automated Analysis of Access Control and Privacy Policies

Anh Truong[1(✉)], Silvio Ranise[2], and Thanh Tung Nguyen[1]

[1] Faculty of Computer Science and Engineering,
Ho Chi Minh City University of Technology, Ho Chi Minh City, Vietnam
anhtt@hcmut.edu.vn
[2] Security and Trust Unit, FBK-Irst, Trento, Italy
ranise@fbk.eu

Abstract. Access Control is becoming increasingly important for today ubiquitous systems. Sophisticated security requirements need to be ensured by authorization policies for increasingly complex and large applications. As a consequence, designers need to understand such policies and ensure that they meet the desired security constraints while administrators must also maintain them so as to comply with the evolving needs of systems and applications. These tasks are greatly complicated by the expressiveness and the dimensions of the authorization policies. It is thus necessary to provide policy designers and administrators with automated analysis techniques that are capable to foresee if, and under what conditions, security properties may be violated. In this paper, we consider this analysis problem in the context of the Role-Based Access Control (RBAC), one of the most widespread access control models. We describe how we design heuristics to enable an analysis tool, called ASASPXL, to scale up to handle large and complex Administrative RBAC policies. We also discuss the capability of applying the techniques inside the tool to the analysis of location-based privacy policies. An extensive experimentation shows that the proposed heuristics play a key role in the success of the analysis tool over the state-of-the-art analysis tools.

1 Introduction

Modern information systems contain sensitive information and resources that need to be protected against unauthorized users who want to steal it. The most important mechanism to prevent this is Access Control [10] which is thus becoming increasingly important for today's ubiquitous systems. In general, access control policies protect the resources of the systems by controlling who has permission to access what objects/resources.

Role-Based Access Control (RBAC) [22] is an important security model for access control that has been widely adopted in real-world applications. In RBAC, access control policies specify which users can be assigned to roles which, in turn, are granted permissions to perform certain operations in the system. Although RBAC simplifies the design and management of access control policies,

© Springer-Verlag GmbH Germany 2017
A. Hameurlain et al. (Eds.): TLDKS XXXVI, LNCS 10720, pp. 142–171, 2017.
https://doi.org/10.1007/978-3-662-56266-6_7

modifications of RBAC policies in complex organizations are difficult and error prone activities due to the limited expressiveness of the basic RBAC model. For this reason, RBAC has been extended in several directions to accommodate various needs arising in the real world such as Administrative RBAC (ARBAC) that supports to control the modification of the policies. The main idea of ARBAC is to provide certain specific users, called administrators, some permissions to execute operations, called administrative actions, to modify the RBAC policies. In fact, permissions to perform administrative actions must be restricted since administrators can only be partially trusted. For instances, some of them may collude to, inadvertently or maliciously, modify the policies (by sequences of administrative actions) so that untrusted users can get sensitive permissions. Thus, automated analysis techniques taking into consideration the effect of all possible sequences of administrative actions to identify the safety issues, i.e. administrative actions generating policies by which a user can acquire permissions that may compromise some security goals, are needed.

Several automated analysis techniques (see, e.g., [3,11,14,15,24,29]) have been developed for solving the user-role reachability problem, an instance of the safety issues, in the ARBAC model. Recently, a tool called ASASPXL [18] has been shown to perform better than the state-of-the-art tools on sets of benchmark problems in [13,24]. The main advantage of the analysis technique inside ASASPXL over the state-of-the-art techniques is that the tool can solve the user-role reachability problem with respect to a finite but unknown number of users in the policies manipulated by the administrative actions. However, ASASPXL does not scale to solve problems in some recently proposed benchmarks in [29]. This is because the so-called state explosion problem has not been handled carefully and thus, prevent ASASPXL to tackle such benchmarks.

In this paper, we show how the model checking techniques underlying the tool ASASPXL can scale up to solve the large and complex problem instances in [29]. The main idea is to try to alleviate the state explosion problem, which is well-known problem in model checking techniques, in the analysis of ARBAC policies. To do this, we study how to design heuristics that try to eliminate as much as possible the administrative actions that do not contribute to the analysis of ARBAC polcies. We also perform an exhaustive experiment to conduct the effectiveness of proposed heuristics and compare the performance of our implementation, namely ASASPXL 2.0, with the state-of-the-art analysis tools. The experimental results show that the proposed heuristics play a key role in the success of our analysis tool over the state-of-the-art ones. Additionally, we discuss the capability of applying the model checking techniques inside ASASPXL to the analysis of location-based privacy policies in the context of the centralized trusted party privacy architecture.

The paper is organized as follows. Section 2 introduces the RBAC, ARBAC models, and the related analysis problems. Section 3 briefly introduces the automated analysis tool ASASPXL and the model checking technique underlying it. The proposed heuristics to enable ASASPXL to scale to solve complex user-role reachability problems are described in Sect. 4. The discussion on the application of the automated analysis tool in the context of privacy policy analysis is in Sect. 5. Section 6 summarizes our experiments and Sect. 7 concludes the paper.

2   The Reachability Problem in ARBAC

In *Role-Based Access Control (RBAC)*, access decisions are based on the roles that individual users have as part of an organization. The process of defining roles is based on a careful analysis of how an organization operates. Permissions are grouped by role name and correspond to various uses of a resource. A permission is restricted to individuals authorized to assume the associated role and represents a unit of control, subject to regulatory constraints within the RBAC model. Roles can have overlapping responsibilities and privileges, i.e. users belonging to different roles may have common permissions. For example, within a hospital, the role of doctor can include operations to perform diagnosis, prescribe medication, and order laboratory tests; the role of nurse can be limited to a strict subset of the permissions assigned to a doctor such as order laboratory tests.

We formalize a RBAC policy as follows:

Definition 1. *A* RBAC policy *is a tuple* $(U, R, P, UA, PA, \succeq)$ *where U is a set of users, R a set of roles, and P a set of permissions. A binary relation* $UA \subseteq U \times R$ *represents a user-role assignment, a binary relation* $PA \subseteq R \times P$ *represents a role-permission assignment, and* $\succeq$ *is a role hierarchy.*

A user-role assignment specifies the roles to which the user has been assigned while a role-permission assignment specifies the permissions that have been granted to the role. A role hierarchy $\succeq$ of the policy is a partial order on R. A tuple $r_1 \succeq r_2$ means that r_1 is *more senior* than r_2 for $r_1, r_2 \in R$, i.e., every permission assigned to r_2 is also available to r_1.

Definition 2. *A user u is an* explicit *member of role r when* $(u, r) \in UA$ *while the user u is an* implicit *member of role r if there exists* $r' \in R$ *such that* $r' \succeq r$ *and* $(u, r') \in UA$.

In RBAC, users are not assigned permissions directly, but only acquire them through the roles they have been assigned:

Definition 3. *A user u has permission p if there exists a role* $r \in R$ *such that* $(r, p) \in PA$ *and u is a (explicit or implicit) member of r.*

Example 1. Consider a RBAC policy describing a department in a university as depicted in Fig. 1. The top-left table is the user-role assignment, the top-right is the role-permission assignment, and the bottom is an example of a role hierarchy (The role at the tail of an arrow is more senior than the one at the head).

Let us consider user *Charlie*: he is an *explicit* member of role *Faculty* because the tuple (*Charlie*, *Faculty*) is in the user-role assignment *UA*. Additionally, role *Faculty* has been assigned to permissions *AssignGrades*, *ReceiveHBenefits*, and

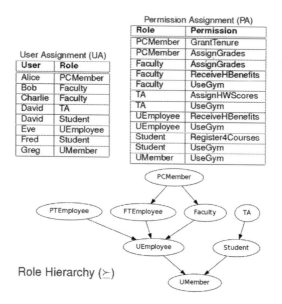

Fig. 1. User and Permission Assignments; and Role Hierarchies

UseGym. Thus, *Charlie* can assign grades, receive benefits and use the gym through the role *Faculty*.

Let us consider the role hierarchy: role *Faculty* is more *senior* than role *UEmployee* (i.e., *Faculty* $\succeq$ *UEmployee*). Therefore, *Charlie* is an *implicit* member of the role *UEmployee*, and thus he can also use all permissions assigned to the role *UEmployee*.   □

2.1   Administrative RBAC (ARBAC)

Access control policies need to be maintained according to the evolving needs of the organization. For instance, the organization promotes some users to new roles, thus, the user-role assignment *UA* of the access control policy needs to be updated (usually by an administrator) according to the change in the organization. For flexibility and scalability in large distributed systems, several administrators are usually required and there is a need not only to have a consistent policy but also to ensure that the policy is modified by administrators who are allowed to do so.

Several administrative frameworks have been proposed on top of the RBAC model to address these issues. One of the most popular administrative frameworks is Administrative RBAC (ARBAC) [8] that controls how RBAC policies may evolve through administrative actions that update the *UA* and *PA* relations (e.g., actions that update *UA* include assigning or revoking user memberships into roles).

Formalization. Usually, administrators may only update the relation UA while PA and $\succeq$ are assumed constant. This is because a change in PA and/or $\succeq$ implies a change in the organization (see [24] for more detail). Thus, we consider ARBAC model that modifies the relation UA while leave PA and $\succeq$ unchanged (e.g., URA 97 [8]). From now on, we focus on situations where U, R, and P are finite, PA and $\succeq$ are constant. Then, a RBAC policy, a tuple $(U, R, P, UA, PA, \succeq)$, can be represented shortly by tuple $(UA, \succeq)$ if U, R, P, and PA are clear from the context ($\succeq$ still appears in the short representation of RBAC policy because the role hierarchy is used to decide the user membership as shown in Definition 2).

Since administrators can be only partially trusted, administration privileges must be limited to selected parts of the RBAC policies, called *administrative domains*. An administrative domain is specified by a *pre-condition* defined as follows:

Definition 4. *A pre-condition C is a finite set of expressions of the forms r or $\bar{r}$ where $r \in R$.*

Then, the relationship between a user and a pre-condition C is defined as follows:

Definition 5. *A user $u \in U$ satisfies a pre-condition C if, for each $\ell \in C$, u is a member of r when ℓ is r or u is not a member of r when ℓ is $\bar{r}$ for $r \in R$. We also say that r is a positive role and $\bar{r}$ is a negative role in C.*

Administrative Actions. RBAC policy $(UA, \succeq)$ can be modified through administrative actions. There are two types of administrative action: permission to assign users to roles is specified by a ternary relation *can_assign*, and permission to revoke users from roles is specified by a binary relation *can_revoke*.

Definition 6. *An administrative action of type can_assign is a tuple of the form (C_a, C, r) where C_a and C are pre-conditions, and r is a role.*

Definition 7. *An administrative action of type can_revoke is a tuple of the form (C_a, r) where C_a is a pre-condition and r is a role.*

In both types, we say that C_a is the *administrative pre-condition*, C is a *(simple) pre-condition*, r is the *target role*, and a user u_a satisfying C_a is the *administrator*. The relation *can_revoke* is only binary because simple pre-conditions are useless when revoking roles (see, e.g., [24]). When there exist users satisfying the administrative and the simple (if the case) pre-conditions of an administrative action, the action is *enabled*.

ARBAC System. An *ARBAC system* is a tuple (α_0, ψ) where α_0 is the *initial* RBAC policy $(UA_0, \succeq)$ and ψ is the (disjoint) union of the following sets of administrative actions of types *can_assign* and *can_revoke* (i.e., $\psi := (can\_assign, can\_revoke)$). A *state* of an ARBAC system is an RBAC policy α. We define the effect of executing an administrative action in ψ by defining a binary relation $\rightarrow_\psi$ on the states of the ARBAC system as follows:

Definition 8. $(UA, \succeq) \rightarrow_\psi (UA', \succeq)$ *iff there exist users u_a and u in U such that either:*

- *there exists $(C_a, C, r) \in can\_assign$, u_a satisfies C_a, u satisfies C (i.e. (C_a, C, r) is enabled), and $UA' = UA \cup \{(u, r)\}$ or*
- *there exists $(C_a, r) \in can\_revoke$, u_a satisfies C_a (i.e. (C_a, r) is enabled), and $UA' = UA \setminus \{(u, r)\}$.*

Clearly, executing an administrative action in ψ changes the ARBAC system status from a certain state to another state. The effect of executing administrative actions in ψ will make a sequence of changes in the states of ARBAC system as follows:

Definition 9. *A run of the ARBAC system $((UA_0, \succeq), \psi)$ is a possibly infinite sequence $(UA_0, \succeq), (UA_1, \succeq), ..., (UA_n, \succeq), ...$ such that $(UA_i, \succeq) \rightarrow_\psi (UA_{i+1}, \succeq)$ for $i \geq 0$.*

Example 2. Consider an ARBAC system (α_0, ψ) where α_0 is the RBAC policy with UA relation and role hierarchy $\succeq$ depicted in Fig. 1, and ψ contains the following administrative action:

$$(\{PCMember\}, \{Student, \overline{TA}\}, PTEmpl) \in can\_assign \qquad (1)$$

The administrative pre-condition is $C_a = \{PCMember\}$, the simple pre-condition is $C = \{Student, \overline{TA}\}$, and the target role is $PTEmpl$.

User *Alice* satisfies the pre-condition C_a because $(Alice, PCMember) \in UA$. User *Fred* satisfies the pre-condition C because he is a *Student* but not a *TA* (e.g., $(Fred, Student) \in UA$ and $(Fred, TA) \notin UA$). As a sequence, the administrative action is enabled.

We now can update the current state $(UA, \succeq)$ of the ARBAC system to a new state $(UA', \succeq)$ where $UA' = UA \cup \{(Fred, PTEmpl)\}$ by executing the following instance of the administrative action (1) specified above: administrator *Alice* (who has role $PCMember$) assigns role $PTEmpl$ to user *Fred* (who is a *Student* and not a *TA*).

Notice that *Alice* cannot assign role $PTEmpl$ to *David* by using the action (1) because he is not only a *Student* but also a *TA* (i.e., *David* does not satisfy the pre-condition C). □

2.2   The User-Role Reachability Problem

Usually, policy designers and administrators want to foresee if the interactions among administrative actions (i.e., runs of the ARBAC system), as seen in the Example 2, can lead the system to conflict states violating the security requirements of the organization (e.g., the security requirements forbid a user to be assigned to some sensitive roles). Thus, they need to analyze access control policies in order to discover such violation. This problem is called as the user-role reachability problem and is defined as follows.

Definition 10. *A* (RBAC) goal *is a pair* (u_g, R_g) *for* $u_g \in U$ *and* R_g *a finite set of roles. The cardinality* $|R_g|$ *of* R_g *is the* size *of the goal.*

Intuitively, a goal can be seen as the negation of a security requirement specified above. Thus, if a goal is reachable in the ARBAC system (i.e., there exists a run of the ARBAC system that leads the system to a state in which the goal is satisfied), the corresponding security requirement is violated.

Definition 11. *Given:*

- *an ARBAC system* (α_0, ψ) *where* α_0 *is the initial RBAC policy* $(UA_0, \succeq)$ *and the set of administrative actions* $\psi = (can\_assign, can\_revoke)$
- *a goal* (u_g, R_g);

(an instance of) the **user-role reachability problem**, *identified by the tuple* $\langle (UA_0, \succeq, \psi), (u_g, R_g) \rangle$, *consists of checking if there exists a finite sequence:*

$$(UA_0, \succeq), (UA_1, \succeq), ..., (UA_n, \succeq)$$

(for $n \geq 0$*) where*

- $(UA_i, \succeq) \rightarrow_\psi (UA_{i+1}, \succeq)$ *for each* $i = 0, ..., n-1$, *and*
- u_g *is a member of each role of* R_g *in* UA_n.

The user-role reachability problem defined here is the same of that considered in [11,13,24]. In case the answer for the user-role reachability problem is *yes*, the goal is **reachable**. Otherwise, the goal is **unreachable**.

Example 3. Consider an ARBAC system (α_0, ψ) where α_0 is the RBAC policy with UA_0 relation and role hierarchy $\succeq$ depicted in Fig. 1, and ψ contains the following administrative actions:

$$(\{PCMember\}, \{Student, \overline{TA}\}, PTEmpl) \in can\_assign \tag{2}$$
$$(\{Faculty\}, \{PTEmpl\}, FTEmpl) \in can\_assign \tag{3}$$

Let us consider the following user-role reachability problem:

$$\langle (UA_0, \succeq, \psi), (Fred, \{FTEmpl\}) \rangle$$

The goal $(Fred, \{FTEmpl\})$ is **reachable** since there exists a finite sequence:

$$(UA_0, \succeq), (UA_1, \succeq), (UA_2, \succeq)$$

where

(i) $UA_1 = UA_0 \cup \{(Fred, PTEmpl)\}$, and
(ii) $UA_2 = UA_1 \cup \{(Fred, FTEmpl)\}$

(i) is done by executing administrative action (2) as shown in Example 2. After this, the current state of the ARBAC system is $(UA_1, \succeq)$. Then, (ii) can be done by executing administrative action (3):

- User Bob satisfies the pre-condition $C_a = \{Faculty\}$ because $(Bob, Faculty) \in UA_1$.
- User $Fred$ satisfies the pre-condition $C = \{PTEmpl\}$ because $(Fred, PTEmpl) \in UA_1$

As a sequence, administrative action (3) is enabled. Thus, we now can update the current state $(UA_1, \succeq)$ to a new state $(UA_2, \succeq)$ where $UA_2 = UA_1 \cup \{(Fred, FTEmpl)\}$ by executing administrative action (3).   □

In real scenario, subtle interactions between administrative actions in real policies may arise that are difficult to be foreseen by policy designers and administrators. Thus, automated analysis techniques are thus of paramount importance to analyze such policies and answer the user-role reachability problem. The analysis techniques we will present in the following will be able to establish this automatically for the problem in ARBAC.

3   Solving the User-Role Reachability Problem

3.1   Model Checking Modulo Theories (MCMT)

MCMT [12] is a framework to solve reachability problems for infinite state systems that can be represented by transition systems whose set of states and transitions are encoded as constraints in a decidable fragment of first-order logic named Bernays-Schönfinkel-Ramsey class. Several systems have been abstracted using such symbolic transition system such as parametrised protocols, sequential programs manipulating arrays, timed system, etc. (see again [12] for an overview).

The main idea inside MCMT framework is to use a backward reachability procedure that repeatedly computes the so-called pre-images of the set of *goal* states of the transition system, that is usually obtained by complementing a certain safety property (e.g., security requirement) that the system should satisfy. Then, the set of backward reachable states of the system is obtained by taking the union of the pre-images. At each iteration of the procedure, the procedure checks whether the intersection between the set of backward reachable states and the initial set of states is non-empty (i.e., *safety* test) or not (i.e., the *unsafety* of the system: there exists a (finite) sequence of transitions that leads the system from an initial state to one satisfying the goal). Otherwise, when the intersection is empty, the procedure checks if the set of backward reachable states is contained in the set computed at the previous iteration (*fix-point* test) and, if yes, the *safety* of the system (i.e. no (finite) sequence of transitions leads the system from an initial state to one satisfying the goal) is returned. Since sets of states and transitions are represented by first-order constraints, the computation of pre-images reduces to simple symbolic manipulations and testing safety and fix-point to solving a particular class of constraint satisfiability problems, called Satisfiability Modulo Theories (SMT) problems, for which scalable and efficient SMT solvers are currently available (e.g., Z3 [2]).

3.2   ASASPXL: An Automated Analysis Tool for ARBAC Policies

In [4,6,17–21], it is studied how the MCMT approach can be used to solve (variants of) the user-role reachability problem. On the theoretical side, it is shown that the backward reachability procedure described above decides (variants of) the user-role reachability problem. On the practical side, extensive experiments have shown that an automated tool, called ASASP [3] implementing (a refinement of) the backward reachability procedure, has a good *trade-off* between *scalability* and *expressiveness*. Immediately after ASASP, a set of much larger instances of the user-role reachability problem has been considered in [13]. Unfortunately, ASASP does not scale to solve the set of problem. This is in line with the following observation of [13]: "model checking does not scale adequately for verifying policies of very large sizes." Then, in [18], a new tool based on the MCMT approach, called ASASPXL, has been proposed to efficiently solve much larger instances of the user-role reachability problem. The new analysis tool ASASPXL is build on top of MCMT, the first implementation of the MCMT approach. The choice of building a new analysis tool instead of modifying ASASP gives some advantage. First, we only need to write a translator from instances of the user-role reachability problem to reachability problems in MCMT input language, a routine programming task. Second, MCMT has been developed and extensively used for the past years. It is thus more robust and offers a high degree of confidence. Third, we can re-use some features of a better engineered incarnation of the MCMT approach that can be exploited to significantly improve performances, as shown in [18].

The structure of ASASPXL is depicted in Fig. 2. It takes as input an instance of the user-role reachability problem without role hierarchy[1] (i.e., the set $\succeq = \emptyset$) and returns **reachable**, when there exists a finite sequence of administrative actions that leads from the initial RBAC policy to one satisfying the goal, and **unreachable** otherwise. To give such results, ASASPXL firstly translates the user-role reachability problem (the initial RBAC policy, the set of administrative actions ψ, and the goal) to the reachability problem in MCMT input language (module **Translator**). Then, it calls the model checker MCMT to verify the reachability of the problem. Finally, according to the answer returned by the model checker (in the data storage **Explored Policies**), ASASPXL refines it and returns **reachable** or **unreachable** as its output (module **Refinement**).

To keep technicalities to a minimum, we illustrate the translation on an instance of the user-role reachability problem as follows.

---

[1] In [18], the authors claim that we can transform a policy with role hierarchies to a policy without them by pre-processing away the role hierarchies as shown in [23]. Then, we only need to process the explicit members of a role when considering the role memberships (cf. Definitions 2 and 5).

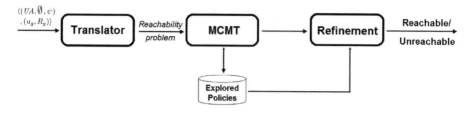

Fig. 2. ASASPXL architecture

Example 4. Let $U = \{u_1, u_2, u_3, u_4, u_5\}$, $R = \{r_1, ..., r_8\}$, initially $UA_0 := \{(u_1, r_1), (u_2, r_2), (u_5, r_5)\}$, and the set ψ contains:

$$(\{r_1\}, \{r_2\}, r_3) \in can\_assign \tag{4}$$

$$(\{r_3\}, \{r_4, \overline{r_5}\}, r_6) \in can\_assign \tag{5}$$

$$(\{r_4\}, \{r_5\}, r_7) \in can\_assign \tag{6}$$

$$(\{r_2\}, \{r_7\}, r_8) \in can\_assign \tag{7}$$

$$(\{r_2\}, r_3) \in can\_revoke \tag{8}$$

$$(\{r_5\}, r_4) \in can\_revoke \tag{9}$$

The goal of the problem is $(u_5, \{r_8\})$.

To formalize this problem instance in MCMT, ASASPXL firstly generates an unary relation u_r per role $r \in R$. The initial relation UA_0 can thus be expressed as

$$\forall x. \left[\begin{array}{l} (u_{r_1}(x) \leftrightarrow x = u_1) \wedge (u_{r_2}(x) \leftrightarrow x = u_2) \wedge (u_{r_5}(x) \leftrightarrow x = u_5) \wedge \neg u_{r_2}(x) \wedge \neg u_{r_3}(x) \wedge \\ \neg u_{r_4}(x) \wedge \neg u_{r_6}(x) \wedge \neg u_{r_7}(x) \wedge \neg u_{r_8}(x) \end{array} \right].$$

where $(u_{r_1}(x) \leftrightarrow x = u_1)$ represents tuple (u_1, r_1) in UA_0; $(u_{r_2}(x) \leftrightarrow x = u_2)$ represents (u_2, r_2) in UA_0 and so on.

An administrative action, for instance, $(\{r_3\}, \{r_4, \overline{r_5}\}, r_6)$ in can_assign is formalized as

$$\exists x \exists y. \left[u_{r_3}(x) \wedge u_{r_4}(y) \wedge \neg u_{r_5}(y) \wedge \forall \lambda. (u'_{r_6}(\lambda) \leftrightarrow (\lambda = y \vee u_{r_6}(\lambda))) \right]$$

and a tuple, for example, $(\{r_2\}, r_3)$ in can_revoke can be expressed as

$$\exists x \exists y. \left[u_{r_2}(x) \wedge u_{r_3}(y) \wedge \forall \lambda. (u'_{r_3}(\lambda) \leftrightarrow (\lambda \neq y \wedge u_{r_3}(\lambda))) \right]$$

where u_r and u'_r indicate the value of U_r immediately before and after, respectively, the execution of the administrative action (we also have omitted—for the sake of compactness—identical updates, i.e. a conjunct $\forall \lambda. (u'_r(\lambda) \leftrightarrow u_r(\lambda))$ for each role r distinct from the target role in the tuple of can_assign or can_revoke). The other administrative actions are translated in a similar way.

The goal $(u_5, \{r_8\})$ can be represented as:

$$\exists x. u_{r_8}(x) \wedge x = u_5$$

The pre-image of the goal, that is computed by the model checker MCMT, with respect to the administrative action $(\{r_2\}, \{r_7\}, r_8)$ is the set of states from which it is possible to reach the goal by using the administrative action $(\{r_2\}, \{r_7\}, r_8)$. This is formalized as the formula (see [6] for details)

$$\exists x \exists y. ((u_{r_7}(y) \wedge y = u_5) \wedge u_{r_2}(x)),$$

On this problem, MCMT returns unreachable (i.e., there does not exist a finite sequence of administrative operations that lead from the initial policy UA to one satisfying the goal). □

Recently, a tool named VAC has been proposed in [11] for solving the user-role reachability problem of ARBAC policies. In [11], it is shown that VAC outperforms RBAC-PAT [24], MOHAWK [13], and even ASASPXL on the problems in [24] and on a new set of complex instances of the user-role reachability problem. It was natural to run ASASPXL on these new benchmark problems: rather disappointingly, it could tackle such problem instances, however, its performance cannot be comparable with the new tool VAC (e.g., ASASPXL returns time-out in some problems). The reason of the bad scalability of ASASPXL is that ASASPXL does not work well on the user-role reachability problems with some specific features such as the problem containing some sub-problems having same structure of administrative actions; and the problems in which no state can be reached from the initial state; etc. These and other problems have lead us to design new heuristics to make ASASPXL more scalable, as we will see in the next sections.

4   ASASPXL 2.0

To enable ASASPXL to scale up to analyze the complex instances of the user-role reachability problem as shown in the previous section, our main idea is to design heuristics that help to alleviate the so-called state explosion problem, one of the commonly known problems in model checking techniques that must be addressed to solve most real-world problems. One of the main sources of complexity is the large number of administrative actions; thus, for scalability, the original set of actions must be refined by using heuristics that tries to eliminate administrative actions that do not contribute to the analysis of RBAC policy. This and other techniques to control the state explosion problem will be detailed in the following (sub-)sections. Before going to the details of heuristics, we emphasize that all heuristics in the following will be implemented in a module named **Heuristics** and will be put before the module **Translator** in the architecture of our implementation, namely ASASPXL 2.0, as shown in Fig. 3. The ASASPXL 2.0's input, a user-role reachability problem, will be processed by module **Heuristics** before being forwarded to module **Translator** and then to module MCMT as described in Sect. 3.

4.1   Backward Useful Actions

The main idea to alleviate the state explosion problem is to eliminate as much as possible administrative actions that is useless to the analysis of ARBAC

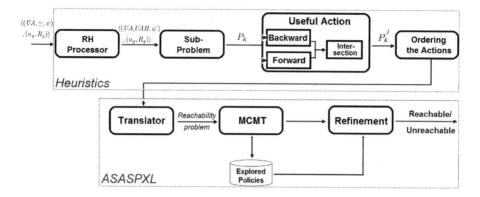

Fig. 3. ASASPXL 2.0: ASASPXL architecture with new heuristics

policy. This is done by extracting increasingly larger sub-sets of the tuples (i.e. administrative actions) in the original set of administrative actions ψ so as to generate a sequence of increasingly more precise approximations of the original instance of the user-role reachability problem. The heuristics to do this is based on the following notion of an administrative action being useful.

Definition 12. *Let ψ be the set of administrative actions and R_g a set of roles in an ARBAC system:*

- *A tuple in ψ is 0-useful iff its target role is in R_g.*
- *A tuple in ψ is k-useful (for $k > 0$) iff it is $(k - 1)$-useful or its target role occurs (possibly negated) in **either** the simple pre-condition **or** the administrative pre-condition of a $(k - 1)$-useful transition.*

A tuple t in ψ is useful iff there exists $k \geq 0$ such that t is k-useful.

The main idea of the proposal of useful actions is that we will find all possible administrative actions in ψ that may contribute (so we say that the actions are useful) to the reachability of R_g. Clearly, the set R_g can be reachable by executing some administrative actions in the set of all 0-useful actions. However, in order to execute some 0-useful administrative actions, we need to execute some 1-useful actions first to satisfy all pre-conditions of the 0-useful actions. Similarly, before executing some 1-useful actions, some 2-useful actions need to be executed first and so on. As a sequence, any k-useful administrative actions may contribute to the reachability of R_g.

Let $\psi^{\leq k} = (can\_assign^{\leq k}, can\_revoke^{\leq k})$ denote the set of all k-useful tuples in $\psi = (can\_assign, can\_revoke)$. It is easy to see that $can\_assign^{\leq k} \subseteq can\_assign^{\leq k+1}$ and $can\_revoke^{\leq k} \subseteq can\_revoke^{\leq k+1}$ (abbreviated by $\psi^{\leq k} \subseteq \psi^{\leq k+1}$) for $k \geq 0$. Since the sets can_assign and can_revoke in ψ are bounded, there must exist a value $\tilde{k} \geq 0$ such that $\psi^{\leq \tilde{k}} = \psi^{\leq \tilde{k}+1}$ (that abbreviates $\psi^{\leq \tilde{k}} \subseteq \psi^{\leq \tilde{k}+1}$ and $\psi^{\leq \tilde{k}+1} \subseteq \psi^{\leq \tilde{k}}$) or, equivalently, $\psi^{\leq \tilde{k}}$ is the (least) fix-point, also denoted with $lfp(\psi)$, of useful tuples in ψ. Indeed, a tuple in ψ is useful iff it is in $lfp(\psi)$.

Example 5. Let ψ be the administrative actions in Example 4 and $R_g := \{r_8\}$. The sets of k-useful tuples for $k \geq 0$ are the following:

$$\psi^{\leq 0} := (\{(\{r_2\}, \{r_7\}, r_8)\}, \emptyset)$$
$$\psi^{\leq 1} := \psi^{\leq 0} \cup (\{(\{r_4\}, \{r_5\}, r_7)\}, \emptyset)$$
$$\psi^{\leq 2} := \psi^{\leq 1} \cup (\emptyset, \{(\{r_5\}, r_4)\})$$
$$\psi^{\leq k} := \psi^{\leq 2} \text{ for } k > 2$$

Now, let us consider the user-role reachability problem: $\langle (UA, \emptyset, \psi^{\leq 2}), (u_5, \{r_8\}) \rangle$. Notice that the role hierarchy $\succeq= \emptyset$ here since ASASPXL can only process user-role reachability problems without role hierarchy as shown in Sect. 3.2. We propose an efficient pre-processing module in Sect. 4.6 that pre-processes away role hierarchies so that (an adapted version of) the technique in ASASPXL can be used to solve user-role reachability problems with role hierarchy (e.g., $\succeq \neq \emptyset$). On this problem, ASASPXL returns unreachable. We obtain the same result if we run the tool on the translation of the following problem instance: $\langle (UA, \emptyset, \psi), (u_5, \{r_8\}) \rangle$ as illustrated in Example 4.   □

This leads to the following proposition.

Proposition 1. *A goal (u_g, R_g) is unreachable from an initial user-role assignment relation UA by using the administrative operations in ψ iff (u_g, R_g) is unreachable from UA by using the administrative operations in $lfp(\psi)$.*

The proof of this fact can be obtained by slightly adapting the proof for the proposition in [18] and is thus omitted here.

4.2 Forward Useful Actions

In Sect. 4.1, we have introduced a heuristic identifying the set of useful actions (that is a subset of the original set of administrative actions) that is enough for solving the user-role reachability. The heuristic initially uses the roles in the goal to identify 0-useful actions and then using roles in the pre-conditions of k-useful actions to decide $(k + 1)$-useful actions. Dually, we can start with the roles in the initial user-role assignment UA_0 and *forwardly* compute the set of useful actions. This is captured by the notion of *forward* useful action as follows:

Definition 13. *Let ψ be the set of administrative actions, R be the set of roles, and $R_i := \{r | (u, r) \in UA_0\} \cup \{\overline{r} | r \in R\}$ be a set of roles occurring in the initial policy UA_0. A tuple $\tau \in \psi$:*

- *is forward 0-useful iff its pre-condition is a subset of R_i*
- *is forward k-useful (for $k > 0$) iff it is:*
  - *$(k - 1)$-useful or,*
  - *its pre-condition is a subset of $R_i = R_i \cup \{r | r$ is the target role of a $(k - 1)$-useful action$\}$*

τ is forward useful iff there exists $k \geq 0$ such that τ is forward k-useful.

Let $\psi_F^{\leq k} = (can\_assign_F^{\leq k}, can\_revoke_F^{\leq k})$ denote the set of forward k-useful actions in $\psi = (can\_assign, can\_revoke)$, it is easy to see that $\psi_F^{\leq k} \subseteq \psi_F^{\leq k+1}$ for $k \geq 0$ and there exists a value $\tilde{k} \geq 0$ such that $\psi_F^{\leq \tilde{k}} = \psi_F^{\leq \tilde{k}+1}$ (i.e., $lfp_F(\psi) = \psi_F^{\leq \tilde{k}}$) since the set ψ is bounded. Similar to the heuristic for backward useful actions above, we conclude the following proposition.

Proposition 2. A goal (u_g, R_g) is unreachable from an initial user-role assignment relation UA by using the administrative operations in ψ iff (u_g, R_g) is unreachable from UA by using the administrative operations in $lfp_F(\psi)$.

Example 6. Let us consider again Example 4. The set R_i of roles in UA_0 is $\{r_1, r_2, r_5, \overline{r_1}, \overline{r_2}, ..., \overline{r_7}, \overline{r_8}\}$.
The sets of forward k-useful tuples for $k \geq 0$ are the following:

$$\psi_F^{\leq 0} := \{(\{r_1\}, \{r_2\}, r_3), (\{r_2\}, r_3), (\{r_5\}, r_4)\}$$
$$\psi_F^{\leq k} := \psi_F^{\leq 0} \text{ for } k > 0,$$

ASASPXL returns `unreachable` on the user-role reachability problem $\langle (UA, \emptyset, \psi_F^{\leq 0}), (u_1, \{r_8\}) \rangle$ that confirms the results in Examples 4 and 5. □

4.3   Integration Backward and Forward Useful Actions in ASASPXL

Combination of Backward and Forward Useful Actions. Clearly, the goal (u_g, R_g) of a user-role reachability problem $\langle (UA_0, \emptyset, \psi), (u_g, R_g) \rangle$ is reachable if there exists a finite sequence of actions in the set $lfp(\psi)$ that leads the ARBAC system from the initial state $(UA_0, \emptyset)$ to a state satisfying the goal (cf. Proposition 1 in Sect. 4.1). Similarly, if the goal is reachable, all the administrative actions in the sequence must belong to the set $lfp_F(\psi)$ (cf. Proposition 2 in Sect. 4.2). This leads to the following idea: compute the intersection of the two sets and then use it for solving the user-role reachability problem. Thus, the combination module, namely **Intersection** in Fig. 3, works as follows to take into consideration the forward and backward useful actions: First, the module computes $lfp(\psi)$ and $lfp_F(\psi)$ that are the set of backward useful and forward useful actions, respectively. Then, the module will compute the intersection ψ_U of the two sets that is expected to be much smaller than $lfp(\psi)$, $lfp_F(\psi)$, and the original set ψ. Finally, the set of useful actions ψ_U is used to replace the original set ψ in solving the user-role reachability problem. The correctness and completeness of taking into consideration the intersection instead of the set of forward or backward useful actions is guaranteed by Proposition 3 that is simply a corollary of Propositions 1 and 2.

Proposition 3. A goal (u_g, R_g) is unreachable from an initial user-role assignment relation UA by using the administrative actions in ψ iff (u_g, R_g) is unreachable from UA by using the administrative operations in $lfp(\psi) \cap lfp_F(\psi)$.

Incremental Analysis Using Backward and Forward Useful Actions.
As mentioned above, one of the main sources of complexity of the analysis is the
large number of administrative actions. The intersection $\psi_U = lfp(\psi) \cap lfp_F(\psi)$
is expected to be much smaller than the original set ψ and as a sequence, the
analysis of the user-role reachability problem with ψ_U is much simpler than
the analysis with ψ. However, to compute ψ_U, we need to compute both $lfp(\psi)$
and $lfp_F(\psi)$ and in some cases, the intersection between them still contains many
administrative actions that may make the analysis still complex. Thus, for better
scalability, the ψ_U should be split into smaller sub-sets that tries to maximize
the probability of ASASPXL to return reachable. The main idea is to extract
increasingly larger sub-sets of the tuples in the original set ψ so as to generate
a sequence of increasingly more precise approximations of the original instance
of the user-role reachability problem. To do this, we iteratively compute $\psi^{\leq k}$
and $\psi_F^{\leq k}$ that are the set of backward k-useful and forward k-useful actions,
respectively, and then, we will compute the intersection $\psi_U^{\leq k} = \psi^{\leq k} \cap \psi_F^{\leq k}$ for
each $k = 0, 1, 2, \ldots$ The set $\psi_U^{\leq k}$ will be used to analyze the user-role reachability
problem. If the answer is reachable, this will be the final answer for the original
problem. Otherwise, we increase k and then repeat the computation. This process
terminates when both $\psi^{\leq k}$ and $\psi_F^{\leq k}$ reach $lfp(\psi)$ and $lfp_F(\psi)$, respectively.

The module **Useful Action** in Fig. 3 uses the idea above to build a
sequence of increasingly precise instances of user-role reachability problem. Such
a sequence is terminated either when the goal is found to be reachable or when
the fix-point of useful administrative actions is detected. Given an instance
$\langle (UA, \emptyset, \psi, (u_g, R_g)) \rangle$ of the user-role reachability problem, the module **Useful
Action** works as follows:

1. Let $k := 0$
2. Compute BUT and FUT, the sets of backward and forward k-useful actions
   in ψ, respectively
3. Compute $IUT = BUT \cap FUT$
4. Repeat
   (a) If IUT is not empty:
       i. **Translate** the instance $\langle (UA, \emptyset, IUT), (u_g, R_g) \rangle$ of the user-role
          reachability problem to MCMT input language
       ii. If MCMT returns reachable, then return reachable
   (b) Let $k := k + 1$, $pBUT := BUT$, and $pFUT := FUT$
   (c) Compute BUT and FUT, the sets of backward and forward k-useful
       actions in ψ, respectively
   (d) Compute $IUT = BUT \cap FUT$
5. Until $pBUT = BUT$ and $pFUT = FUT$
6. Return unreachable

Initially, BUT contains $\psi^{\leq 0}$, FUT contains $\psi_F^{\leq 0}$, and $IUT = \psi^{\leq 0} \cap \psi_F^{\leq 0}$ (steps
2 and 3). At iteration $k \geq 1$, BUT stores $\psi^{\leq k}$ and $pBUT$ contains $\psi^{\leq (k-1)}$.
Similarly, FUT stores $\psi_F^{\leq k}$ and $pFUT$ contains $\psi_F^{\leq (k-1)}$. For $k \geq 0$, the instance
$\langle (UA, \emptyset, IUT), (u_g, R_g) \rangle$ of the user-role reachability problem is translated to

MCMT input language (step 4(a)(i)). In case MCMT discovers that the goal (u_g, R_g) is reachable with the sub-set IUT of the administrative actions, *a fortiori* (u_g, R_g) is reachable with respect to the whole set ψ, and the module returns (step 4(a)(ii)). Otherwise, a new instance of the user-role reachability problem is considered at the next iteration if the condition at step 5 does not hold, i.e. BUT and FUT do not yet store $lfp(\psi)$ and $lfp_F(\psi)$, respectively. If the condition at step 5 holds, by Proposition 3, we can exit the loop and return the unreachability of the goal with respect to the whole set ψ of administrative actions. The termination of the loop is guaranteed by the existence of $lfp(\psi)$ and $lfp_F(\psi)$.

4.4   Ordering Administrative Actions

We recall that the module MCMT implements the backward reachability procedure that computes the sets of backward reachable states from the goal. Basically, at each iteration, the procedure takes the first administrative action in the set ψ, computes its backward reachable states (pre-image) and then checks the intersection between the initial state and the backward states (by using an SMT solver to check the satisfiability). If the intersection is not empty (i.e., the goal is reachable from the initial state), the procedure returns `reachable` and stops. Otherwise, it selects the second action and repeats the process until all actions have been considered. This idea gives two advantages: first, the procedure can stop as soon as possible when it decides that the goal is reachable by checking an action and thus, not necessary to check the remaining actions; second, the fix-point formula can be divided into a set of smaller formulae, namely *local fix-points*, that is easier to be checked by SMT solvers. The original fix-point is reached when all the local fix-points are reached.

Clearly, the selection of the next action for computing the pre-images should be handled carefully since this will cause some redundant in the analysis that may negatively affect the performances of the procedure. In fact, if the goal is reachable and the administrative action, let say τ, that helps the procedure in deciding the reachability of the goal is at the end of the action list, the current version of the backward reachability procedure must computes the pre-images for all actions before τ that are actually redundant computations. It is thus desirable to design a heuristics to select the next action to maximize the possibility of picking up an action that is important to show the reachability of the goal.

Our heuristics is based on the idea of how "close" between the set of states produced by computing the pre-image with respect to a given action and the set of initial states. This is because for each iteration, the procedure checks if the intersection between the pre-image generated by the given action and the set of initial states is empty, and then uses this check to decide the reachability of the goal. To illustrate how an action is "closer" than another, let us consider the following example:

Example 7. Let $U = \{u_1, u_2\}$, $R = \{r_a, r_1, ..., r_7\}$ initially $UA := \{(u_1, r_a), (u_1, r_1), (u_1, r_2), (u_1, r_5)\}$, and the set ψ contains:

$$(\{r_a\}, \{r_1, r_2, \overline{r_4}\}, r_7) \in can\_assign \tag{10}$$

$$(\{r_a\}, \{r_1, r_3\}, r_7) \in can\_assign \tag{11}$$

The pre-images of the two actions (10) and (11) (computed by the backward reachability procedure) are represented by formulae $\exists x, y.(r_a(x) \wedge r_1(y) \wedge r_2(y) \wedge \neg r_4(y))$ and $\exists x, y.(r_a(x) \wedge r_1(y) \wedge r_3(y))$, respectively. It is easy to see that the set of reachable states of action (10) is contained in the initial state UA (i.e., their intersection is not empty). We also notice how all the roles in the precondition of action (10) appear in UA while role r_3 in the precondition of action (11) does not. In this case, we say that action (10) is closer (to the initial state) than action (11). Then, action (10) should be selected before action (11) in the backward reachability procedure. □

We define the function *Diff* calculating how "close" two sets of roles are as follows:

Definition 14. *Let C_1 and C_2 be pre-conditions, the difference between C_1 and C_2 is:*

$$Diff(C_1, C_2) := (C_1^+ \setminus C_2^+) \cup (C_1^- \setminus C_2^-)$$

where C_1^+ and C_2^+ are sets of positive roles in C_1 and C_2, respectively; C_1^- and C_2^- are sets of negative roles in C_1 and C_2, respectively.

We illustrate how the function *Diff* is used in the heuristic by the following example:

Example 8. Let us consider again Example 7. First, the heuristic will calculate $R_i = \{r_a, r_1, r_2, r_5, \overline{r_a}, \overline{r_1}, ..., \overline{r_7}\}$ that represents all roles occurring in the initial UA as defined in Definition 13.

Let consider action (10) with its precondition $C_1 = \{r_a, r_1, r_2, \overline{r_4}\}$, the heuristic then computes $Diff(C_1, R_i) = \emptyset$. Similarly, the precondition of action (11) is $C_2 = \{r_a, r_1, r_3\}$ and $Diff(C_2, R_i) = \{r_3\}$.

Since $|Diff(C_2, R_i)| > |Diff(C_1, R_i)|$, we say that action (10) is closer (to the initial state) than action (11). In other words, the precondition C_1 can be easily satisfied by the initial UA while C_2 requires more tuples, for instance $(u_1, r_3) \in UA$, to be satisfied. Thus, the heuristic will select the actions (10) to compute its pre-image before (11). □

The module **Ordering the Actions** in Fig. 3 implements the heuristic mentioned above. After computing the set of useful actions as in Sect. 4.3, the module **Ordering the Actions** will be invoked with the set ψ_U^k of useful actions as the parameter. The module then orders the administrative actions in ψ_U^k and returns the ordered set as shown in the following workflow:

1. Let ψ_U be the set of actions and R_i containing all roles occurring in the initial RBAC policy UA_0

2. For each $\tau = (C_a, C, r) \in \psi_U^k$:
   (a) If $C_a = \emptyset$ and $C = \emptyset$:
       i. set τ be the first order in ψ_U^k (for several actions with $C_a = C = \emptyset$, we do not care the order between them)
   (b) Else:
       i. Calculate $Diff_\tau := Diff(C_a \cup C, R_i)$ for τ
3. Order the actions in ψ_U^k by their $|Diff_\tau|$ (from lower value to higher one)
   (a) If $|Diff_{\tau_1}| = |Diff_{\tau_2}|$ where $\tau_1 = (C_{a1}, C_1, r_1)$ and $\tau_2 = (C_{a2}, C_2, r_2)$:
       i. τ_1 has higher order if $|C_{a1} \cup C_1| < |C_{a2} \cup C_2|$ and vice versa

Initially, the procedure computes the set R_i containing all roles in the initial UA_0. Then, it calculates the set $Diff$ for each administrative action in ψ_U^k (Step 2). Administrative actions of the form $(\emptyset, \emptyset, r)$ are set highest order in Step 2(a) since its pre-conditions are alway satisfied. For several actions of the form $(\emptyset, \emptyset, r)$, we do not care about the order between them. The procedure then classifies the actions in ψ_U based on their $Diff$ (Step 3). Notice how the procedure prioritizes the action containing smaller set of pre-conditions (Step 3(a)) for the actions having the same $|Diff|$. This is because the formula representing the set of backward reachable states generated by the action (see, e.g., Example 7) may be smaller (i.e., containing less literals) than the others and thus easier for the SMT solver to check the satisfiability.

4.5   Sub-problems

The size of the goal (i.e., $|R_g|$, for a user-role reachability problem $\langle(UA, \succeq, \psi), (u_g, R_g)\rangle$) is also one of the main sources of the complexity of the problem (e.g., see [24]). Indeed, experiments with a preliminary version of our techniques confirmed this result: it did not scale up even to modest values of $|R_g|$. To overcome this situation, we design a *divide et impera* heuristic based on the observation: the goal (u_g, R_g) is reachable if there exists a state $(UA_n, \succeq)$ in a run of ARBAC system such that u_g is a member of *all* roles in R_g (cf. Definition 11). This observation enables us to divide the original problem $\langle UA, \psi, (u_g, R_g)\rangle$ to a set of (sub-)reachability problems $SP = \{\langle(UA, \succeq, \psi), (u_g, r_g)\rangle | r_g \in R_g\}$. Indeed, solving each reachability problem in the set SP is much easier than the original problem and, in many cases, the answers for reachability problems in SP can lead to the final answer for the original system.

The module **Sub-Problem** in Fig. 3 implements the idea of divide et impera heuristic mentioned above. The module is put before the module **Useful Actions**. It inputs the original user-role reachability problem and then splits the problem into a set of sub-problems according to the set $|R_g|$. The module works as follows:

1. Let $P = \langle(UA, \succeq, \psi), (u_g, R_g)\rangle$ be the original problem, $SP = \{\langle(UA, \succeq, \psi), (u_g, r_g)\rangle | r_g \in R_g\}$, and $\sigma = \emptyset$.
2. For each $P_k \in SP$:
   (a) $(Rslt, \sigma_k) =$ call ASASPXL on the problem P_k

(b) $\sigma = \sigma \cup \sigma_k$

(c) If $Rslt = $ unreachable:

    i. Return unreachable on the original problem P.

3. $(Rslt, \sigma) = $ call ASASPXL on the problem $P_\sigma = \langle (UA, \succeq, \sigma), (u_g, R_g) \rangle$

4. If $Rslt = $ reachable:

    (a) Return reachable on the original problem P.

5. Else:

    (a) $(Rslt, \sigma) = $ call ASASPXL on the original problem P

    (b) Return $Rslt$.

Initially, the heuristic computes the set SP containing all sub-problems of the original problem P. Then, it invokes ASASPXL on each sub-problem $P_k \in SP$ (e.g., see Step 2(a)). If there exists a sub-problem $P_k = \langle (UA, \succeq, \psi), (u_g, r_k) \rangle$ for which the result returned by the tool is unreachable, the heuristic immediately returns unreachable for the original problem (Step 2(c)(i)). This is because we cannot find a state $(UA_n, \succeq)$ where $(UA, \succeq) \rightarrow_\psi (UA_n, \succeq)$ such that u_g is a member of r_k and so the goal (u_g, R_g) is never satisfied. We notice how the set σ is enlarged by adding the sequence of actions returned by ASASPXL for each P_k (Step 2(b)).

When all the results of the sub-problems are reachable, we need to invoke again ASASPXL on the problem similar to the original P except the set ψ is replaced by the set σ, which is (hopefully) much smaller than ψ (e.g., see, Step 3). The aim of this heuristics is to check if the original goal is also satisfied (Step 4) by applying only the actions in σ. (Notice that even if all the sub-goals (u_g, r_k) are reachable, there is no guarantee the compositional goal (u_g, R_g) is so by using actions in σ). In the worst case, we invoke ASASPXL on the original problem P (Step 5).

4.6 Solving the User-Role Reachability Problem with Role Hierarchy

Currently, ASASPXL only supports to analyze ARBAC policy without role hierarchy and assumes that the role hierarchy can be pre-processed by the approach proposed in [24] (As the result, the role membership considers only explicit users instead of both implicit and explicit users). The main idea of the pre-processing stage is to replace each action in the original set of administrative actions ψ by a set of additional actions with respect to the hierarchies of roles being present in the pre-conditions of the original action. However, as shown in [21], this pre-processing stage results in an *exponential number* of additional administrative actions in ψ. Clearly, the number of administrative actions is the main source of complexity in solving the user-role reachability problem. As a sequence, the analysis technique solving the user-role reachability problems with role hierarchy by applying the pre-processing approach in [24] may not scalable. In this section, we apply a new approach proposed in [21,26] in processing role hierarchies. The approach is based on the following crucial observation: if a user is assigned to a role r_1 and $r_1 \succeq r_2$, we can assume that the user is also assigned to role r_2.

In this case, we say that the user is implicitly assigned to role r_2. This suggests to transform each tuple in the role hierarchy $\succeq$ to a new administrative action of type *can_assign_hier* (similar to those of type *can_assign*) such that when a user is assigned to a role r, he can be "implicitly" assigned to any junior role of r by executing the new actions. As we need only one additional action per tuple in role hierarchy $\succeq$, it is easy to see that the number of such actions is linear in the cardinality of the role hierarchy $\succeq$.

The effect of explicit and implicit role memberships must be handled carefully. In fact, if a user u is assigned to a role r by a *can_assign* action (explicit role membership), u then can be implicitly assigned to any junior role of r by executing the *can_assign_hier* actions mentioned above (implicit role membership). Now, if u is revoked from r by executing a *can_revoke* action, then the role membership must be handled in a way such that also all the junior roles of r that have been implicitly assigned to u must be revoked. Intuitively, to do this, we need to keep track of all the junior roles implicitly assigned to every user that is a computationally heavy task. To avoid this problem, we modify the structure of RBAC policies defined in Sect. 2 by adding a new relation $UAH \subseteq U \times R$. Now, the relation UA is required to record only explicit role memberships (i.e., those resulting by executing *can_assign* actions) while the new relation UAH record both the explicit and implicit ones.

Before describing the approach to process the role hierarchy, we introduce the new administrative action of type *can_assign_hier* as follows.

Definition 15. *An administrative action of type* can_assign_hier *is of the form* $(r_s \diamond r_j)$ *where* r_s *and* r_j *are roles in* R.

Moreover, to handle the effect of explicit and implicit role memberships, we need to modify the relation $\rightarrow_\psi$ defined in Definition 8 in Sect. 2.1 as in the following. Notice that a state of ARBAC system is modified by adding the new relation UAH and removing relation $\succeq$ (since all tuples in $\succeq$ are transformed to *can_assign_hier* actions):

Definition 16. $(UA, UAH) \rightarrow_\psi (UA', UAH')$ *iff there exist users* u_a *and* u *in* U *such that either:*

- *there exists* $(C_a, C, r) \in$ *can_assign,* u_a *satisfies* C_a, *u satisfies* C *(i.e.* (C_a, C, r) *is enabled),* $UA' = UA \cup \{(u,r)\}$, *and* $UAH' = UAH \cup \{(u,r)\}$ *or*
- *there exists* $(C_a, r) \in$ *can_revoke,* u_a *satisfies* C_a *(i.e.* (C_a, r) *is enabled),* $UA' = UA \setminus \{(u,r)\}$, *and* $UAH' = UA'$ *or*
- *there exists* $(r_s \diamond r_j) \in$ *can_assign_hier,* u *satisfies* $\{r_s\}$, *and* $UAH' = UAH \cup \{(u,r_j)\}$.

We note that the satisfiability of a user to a pre-condition is now with respect to relation UAH instead of UA as in Sect. 2 (cf. Definition 2). This is because we need to consider both implicit and explicit role memberships. We also emphasize that *can_assign* actions update both UA and UAH while *can_assign_hier* ones update only UAH. Additionally, *can_assign_hier* actions do not require an administrator to be executed, it only requires to check that there exists a user u who is member of senior role r_s to add the tuple (u, r_j) to UAH. An action of type can_revoke removes a tuple from UA and then sets UAH to the updated value (i.e., after the removal of the tuple) of UA. The need of resetting UAH to UA after removing a tuple arises from the observation that removing an explicit role membership invalidates all the implicit ones in UAH related to it.

The module **RH Processor** in Fig. 3 implements the approach to pre-process the role hierarchy. The module will input the original user-role reachability problem with role hierarchy. Then, it pre-processes away the role hierarchy by transforming all tuple in the role hierarchy into a set of can_assign_hier actions. The module works as follows (Note: $\overline{Senior(r)}$ stands for a set of all senior roles of r with respect to hierarchy $\succeq$ but is written in negative form $\overline{r_i}$ where r_i is a senior role of r):

1. Processing negative roles in pre-conditions:
    (a) For each tuple $(C_a, C, r) \in can\_assign$:
        i. for each negative role $\overline{r}$ occurring in C_a: replace $\overline{r}$ with $\overline{Senior(r)}$ with respect to role hierarchy $\succeq$.
        ii. for each negative role $\overline{r}$ occurring in C: replace $\overline{r}$ with $\overline{Senior(r)}$ with respect to role hierarchy $\succeq$.
    (b) For each tuple $(C_a, r) \in can\_revoke$:
        i. for each negative role $\overline{r}$ occurring in C_a: replace $\overline{r}$ with $\overline{Senior(r)}$ with respect to role hierarchy $\succeq$
2. Processing tuples in role hierarchy $\succeq$:
    (a) For each tuple $(r_s \succeq r_j) \in \succeq$:
        i. add a can_assign_hier action $(r_s \diamond r_j)$ to the set of administrative actions ψ

Clearly, the number of additional actions resulting by applying this approach is linear to the number of tuples in the role hierarchy $\succeq$ since each tuple in the role hierarchy is transformed to exactly a can_assign_hier action.

Example 9. Consider the ARBAC system in Example 2 and the role hierarchy $\succeq$ as in Fig. 1. Using step 1, the administrative action $(\{PCMember\}, \{Student, \overline{TA}\}, PTEmpl) \in can\_assign$ will be transformed to the following action:

$$(\{PCMember\}, \{Student, \overline{TA}\}, PTEmpl) \in can\_assign \tag{12}$$

because TA does not have any senior role according to the role hierarchy.

Then, the module **RH Processor** adds to the set ψ the following can_assign_hier actions:

$$(PCMember \diamond FTEmployee), \tag{13}$$
$$(PCMember \diamond Faculty), \tag{14}$$
$$(PTEmployee \diamond UEmployee), \tag{15}$$

$$(FTEmployee \diamond UEmployee), \tag{16}$$

$$(Faculty \diamond UEmployee), \tag{17}$$

$$(UEmployee \diamond UMember), \tag{18}$$

$$(TA \diamond Student), \tag{19}$$

$$(Student \diamond UMember), \tag{20}$$

□

Finally, we emphasize that the suggestion of the new type of action *can_assign _hier* does not effect the heuristics proposed in Sects. 4.1, 4.2, 4.3, 4.4, and 4.5. Indeed, for all the heuristics, we only need to process *can_assign_hier* actions as in the way like *can_assign* actions.

4.7   Putting Things Together

We now describe the flow of execution among the various modules in ASASPXL 2.0 (see Fig. 3). The input instance $P = \langle (UA, \succeq, \psi), (u_g, R_g) \rangle$ of the user-role reachability problem is given to the **RH Processor** module that pre-processes the role hierarchy as shown in Sect. 4.6. The original problem is then transformed to a new problem $P = \langle (UA, UAH, \psi'), (u_g, R_g) \rangle$. Next, the new problem is forwarded to the **Sub-Problem** module that splits the new problem into a set of sub-problems P_k by using the heuristic mentioned in Sect. 4.5. Each sub-problem P_k is then processed by the **Useful Action** module to eliminate administrative actions in ψ' that do not contribute to the analysis of the ARBAC system (see Sect. 4.3). At this step, we use the incremental analysis as shown in Sect. 4.3, thus, each sub-problem of P_k, namely P_k^j, will be then forwarded to the **Ordering the Actions** module for sorting the order of administrative actions. For the next steps, ASASPXL will be invoked and the flow of execution is as shown in Sect. 3.2.

5   ASASPXL 2.0 in the Analysis of Location-Based Privacy Policy

The rapid development of location-based services (LBS), which make use of the location information of users, gives both opportunities and challenges for users and service providers [5,9]. The opportunities are that users can get benefits from the services while the service providers can earn more profits. However, by using services, users face with location privacy problem because his location data is attractive to attackers. For example, from the location data of a user, the attackers can decide who the user is, where the user is, what the user's hobbies are, and so on. Thus, it requires a number of solutions to protect the location privacy of users while not affecting much on the quality of the location-based services [25,27,28].

Indeed, the user location privacy should be protected at the time of using LBS services. One popular approach is to obfuscate the user's real location, i.e., the

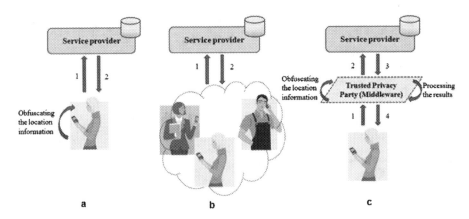

Fig. 4. Privacy architectures: the non-cooperative architecture (a); the peer-to-peer cooperative architecture (b); the centralized trusted party architecture (c)

real position of the user is blurred to decrease the accuracy. There are numerous techniques to obfuscate a position such as enlarging the area containing the position or shifting the real position [5].

There are three popular system architectures for preserving location privacy (Fig. 4): the non-cooperative architecture, the peer-to-peer cooperative architecture, and the centralized trusted party architecture [16,28]. In the non-cooperative architecture, users are self-responsible for protecting their location privacy. For example, the users can create many dummy locations and then send them to the service providers instead of his real location. This is an easy way to protect location privacy but the critical foible is that it totally depends on users' knowledge. In the peer-to-peer cooperative architecture, users are gathered into groups and then a random user in the group acts as the deputy to send requests of the group's users to service providers. By this way, it is difficult to identify exactly the user who really issues the service request [16]. The main problem of this approach is that sometimes a user cannot form a group since there is not enough surrounding cooperative users.

The most widely-used approach to preserve location privacy is the centralized trusted party architecture [28]. In this approach, a trusted party stands between the users and the service providers that provides location transparency mechanism [7]. When using a service, the user sends his location information to the trusted party and then the party blurs the location information using transparency mechanisms. The party then sends the blurred location to the service providers. Finally, it receives results returned from the service providers, filters the results, and forwards back the results to the user. This architecture fulfills the weak point of the first architecture because it does not rely on users awareness. Moreover, the architecture is flexible as it separates functional module (the service providers) and the privacy module (the trusted party). However,

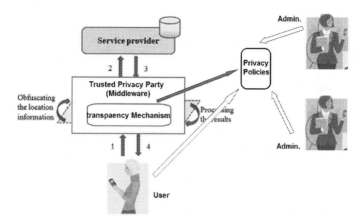

Fig. 5. Privacy preserving module in the centralized trusted party architecture

the disadvantages in this architecture are bottleneck problem and how trust the third party is [16].

In order to preserve location privacy with respect to user's requirement, privacy policies should be set up and put inside the trusted party (Fig. 5). In general, a privacy policy specifies which level of privacy according to user's requirement [27]. A privacy level describes how much the real location of a the user is blurred using transparency mechanisms. Each user may have some different requirements of privacy level. For instance, when the user is at home, he may require higher privacy level than at working place. In a large or distributed trusted party, privacy policies can be changed by the user or by some administrators according to the changes in the user's requirement or in management policies of the trusted party. This may lead the trusted party system to a conflict state in which the privacy of a user is violated since the interaction between the changes of the user and administrators (for example, the privacy level of a user is inadvertently or maliciously reduced by the combination of user's and administrators' actions to the privacy policies). Thus, it needs some analysis techniques that analyse the changes made by users and administrators to verify whether a conflict state happens in the privacy policies of the trusted party or not.

Intuitively, the analysis technique inside ASASPXL 2.0 can help to analyse the privacy policies in the context of centralized trusted party. To do this, we need to specify the trusted party system clearly, i.e., the structure of a privacy policy, the components of administrative actions that are used by users and administrators to modify the privacy policies, and the definition of conflict states. Moreover, such trusted party system must be designed in such a way that the analysis procedure inside the technique in ASASPXL 2.0 terminates (see [20] for more details). We leave the specification of the system and the details of the privacy policy analysis based on the techniques inside ASASPXL 2.0 as our future work.

6   Experiments

We have implemented ASASPXL 2.0 and heuristics in Python and used the MCMT model checker [1] for computing the pre-images. We have also conducted an experimental evaluation to show the scalability of ASASPXL 2.0 and compare it with state-of-the-art analysis tools such as MOHAWK [13], VAC [11], and PMS [29] on two benchmark sets from [11,13]. Note that PMS contains 2 versions, namely *Prl* and *Fwd* that implement the analysis with/without applying their parallel algorithm (see [29] for more details).

Remark. Sometimes, to simplify the analysis of ARBAC policies, *separate administration assumption* (for short, SA) has been applied (see, e.g. [24]) which amounts to requiring that administrative roles (i.e., roles occurring in the administrative precondition C_a) and regular roles (i.e., roles occurring in the simple precondition C) are disjoint. This permits to consider just one user, omit administrative users and roles so that the tuples in *can_assign* are pairs composed of a simple precondition and a target role (i.e., (C, r)) and the pairs in *can_revoke* reduce to target roles only (i.e., (r)). In the state-of-the-art analysis tools mentioned above, MOHAWK requires this assumption while the other two and ASASPXL 2.0 do not need it. The benchmarks are thus classified as either SA benchmarks (that require SA assumption) or non-SA benchmarks (that do not require the assumption) as in the following.

Description of Benchmarks. The first benchmark set is a SA benchmark taken from [13]. It contains three synthetic test suites: **Test suite 1** contains policies in which roles occur only positively in the (simple) pre-conditions of *can_assign* rules and the set of *can_revoke* rules is non-empty. **Test suite 2** contains policies in which roles occur both positively and negatively in *can_assign* rules and the set of *can_revoke* rules is empty. **Test suite 3** contains policies in which roles occur both positively and negatively in *can_assign* rules and the set of *can_revoke* rules is non-empty. The second benchmark set is a non-SA benchmark from [29]. It contains 10 instances of the user-role reachability problem inspired by a university.

Evaluation. We perform all the experiments on an Intel Core I5 (2.6 GHz) CPU with 4 GB Ram running Ubuntu 11.10.

Table 1 reports the results of running ASASPXL 2.0, PMS, VAC and MOHAWK on the first benchmark set. Notice that all problems in this benchmark are unsafe (i.e., analysis tools returns "reachable"). Column 1 shows the name of the test suite, column 2 contains the number of roles and administrative operations in the policy. Columns 3, 4, 6 and 7, and 8 show the average times (in seconds) taken by MOHAWK, VAC, PMS (with two versions), and ASASPXL 2.0, respectively, to solve the instances of the user-role reachability problem associated to an ARBAC policy. For MOHAWK and VAC, the average times also include the time spent in the slicing phase (a technique for eliminating irrelevant users, roles, and administrative operations that are non relevant to solve a certain instance of the

Table 1. Experimental results on the "complex" benchmarks in [13](Separate Administration Assumption)

Test suite	# Roles ◇ # Rules	MOHAWK	VAC		PMS		ASASPXL 2.0	
					Fwd	*Prll*		
		Time	Time	# Rules	Time	Time	Time	# Rules
Test suite 1	3 ◇ 15	0.45	0.29	1	0.38	0.45	**0.09**	1
	5 ◇ 25	0.53	0.35	1	0.38	0.47	**0.10**	1
	20 ◇ 100	0.64	0.35	1	0.35	0.39	**0.13**	1
	40 ◇ 200	0.97	0.69	1	0.49	0.57	**0.36**	1
	200 ◇ 1000	2.69	0.95	1	0.47	0.55	**0.33**	1
	500 ◇ 2500	4.88	1.59	1	0.97	1.16	**0.70**	1
	4000 ◇ 20000	16.99	1.88	1	33.55	22.39	**1.27**	1
	20000 ◇ 80000	51.57	2.72	1	*TO*	*TO*	**1.25**	2
	30000 ◇ 120000	65.51	4.12	1	*TO*	*TO*	**1.65**	1
	40000 ◇ 200000	131.17	9.94	1	*TO*	*TO*	**2.21**	2
Test suite 2	3 ◇ 15	0.45	0.25	1	0.36	0.37	**0.15**	1
	5 ◇ 25	0.55	0.39	1	0.35	0.38	**0.27**	1
	20 ◇ 100	0.59	0.24	1	0.32	0.49	**0.19**	1
	40 ◇ 200	1.21	0.56	1	0.54	0.59	**0.16**	1
	200 ◇ 1000	2.55	0.83	1	0.59	0.63	**0.21**	1
	500 ◇ 2500	6.12	1.52	1	1.54	0.83	**0.49**	2
	4000 ◇ 20000	15.51	1.63	1	29.17	21.39	**1.19**	1
	20000 ◇ 80000	26.12	5.25	1	*TO*	*TO*	**1.20**	2
	30000 ◇ 120000	98.95	6.73	1	*TO*	*TO*	**1.25**	1
	40000 ◇ 200000	146.84	11.89	1	*TO*	*TO*	**1.47**	2
Test suite 3	3 ◇ 15	0.51	0.15	1	0.37	0.35	**0.09**	1
	5 ◇ 25	0.45	0.19	1	0.55	0.49	**0.09**	1
	20 ◇ 100	0.87	0.31	1	0.42	0.62	**0.17**	1
	40 ◇ 200	0.99	0.67	1	0.46	0.57	**0.19**	2
	200 ◇ 1000	7.23	2.12	1	0.92	1.28	**0.59**	2
	500, 2500	4.69	1.20	1	0.74	0.97	**0.15**	1
	4000 ◇ 20000	15.15	4.61	1	20.49	15.13	**1.27**	1
	20000 ◇ 80000	32.35	3.85	1	*TO*	*TO*	**2.21**	1
	30000 ◇ 120000	115.11	9.65	1	*TO*	*TO*	**1.65**	2
	40000 ◇ 200000	157.35	10.32	1	*TO*	*TO*	**2.65**	2

TO: time out *Err*: Error *m*: minute

user-role reachability problem, see [11,13] for more details) and the verification phase. Column 5 and 9 represent the number of actions remaining after the slicing phase of VAC and the useful actions obtained by ASASPXL 2.0, respectively.

Experiments for the benchmark that does not adopt the separate administration assumption are reported in Table 2; their columns have the same semantics as in previous table with additional column "Answer" reports the results returned by analysis tools (*Safe* means the goal is unreachable while *Unsafe*

Table 2. Experimental results on the benchmarks in [29](Non Separate Administration Assumption)

Test case	# Roles ◇ # Rules	Answer	VAC		PMS		ASASPXL 2.0	
					Fwd	Prll		
			Time	# Rules	Time	Time	Time	# Rules
Test 1	40 ◇ 487	Unsafe	17.25	3	0.83	**0.68**	1.15	2
Test 2	40 ◇ 450	Safe	0.21	0	0.91	0.75	**0.19**	0
Test 3	40 ◇ 462	Unsafe	9.33	3	0.92	0.93	**0.71**	2
Test 4	40 ◇ 446	Unsafe	7.51	3	0.99	45.16	**0.69**	1
Test 5	40 ◇ 480	Unsafe	48.31	47	1.25	**0.91**	1.72	7
Test 6	40 ◇ 479	Unsafe	26.62	13	1.02	**0.86**	1.59	3
Test 7	40 ◇ 467	Unsafe	1 m12.56	101	4.22	3.26	**1.85**	2
Test 8	40 ◇ 484	Unsafe	1 m16.23	65	5.08	2 m16.21	**2.04**	7
Test 9	40 ◇ 463	Unsafe	1 m35.11	89	5.91	6 m35.24	**2.91**	9
Test 10	40 ◇ 481	Unsafe	29.94	38	**0.65**	0.75	2.15	5

means the goal is reachable). We do not report the experimental result of MOHAWK because it cannot handle user-role reachability problems without the separate administration assumption.

The results clearly show that ASASPXL 2.0 performs significantly better than MOHAWK, PMS, and VAC in the first benchmark set (Table 1). Notice that PMS throws a time-out (that is set to 10 min) in the biggest test cases. For the second benchmark set, ASASPXL 2.0 outperforms PMS and is much better than VAC. We emphasize that the number of actions after using module **Useful Action** in ASASPXL 2.0 is reduced significantly (column 9).

Table 3. Experimental results when turning on/off heuristics in Sect. 4

Test case	# Roles ◇ # Rules	Answer	ASASPXL 2.0	
			Without heuristic	With heuristics
Test 1	40 ◇ 487	Unsafe	2 m52.73	**1.15**
Test 2	40 ◇ 450	Safe	16.22	**0.19**
Test 3	40 ◇ 462	Unsafe	1 m1.63	**0.71**
Test 4	40 ◇ 446	Unsafe	57.15	**0.69**
Test 5	40 ◇ 480	Unsafe	2 m35.87	**1.72**
Test 6	40 ◇ 479	Unsafe	2 m45.71	**1.59**
Test 7	40 ◇ 467	Unsafe	3 m17.33	**1.85**
Test 8	40 ◇ 484	Unsafe	TO	**2.04**
Test 9	40 ◇ 463	Unsafe	TO	**2.91**

TO: time out *Err*: Error *m*: minute

Table 3 shows experimental results when we run ASASPXL 2.0 on the instances of user-role reachability problem in Table 2 with and without heuristics introduced in Sect. 4. Columns 1, 2, and 3 have the same semantic as previous tables. Column 4 reports the analysis time when turning off all heuristics while column 5 shows the performance obtained by using heuristics. The results prove the effectiveness of heuristics on the analysis. In many cases, the analysis time is reduced significantly, for example, from 3 min to nearly 2 s as in Test 7.

7   Conclusions

We have presented techniques to enable the model checking technique inside a tool, called MCMT, to solve instances of user-role reachability problem. One of the commonly known problems in model checking techniques that must be addressed to solve most real-world problems is the state space explosion problem. Thus, for scalability, we have also designed a set of heuristics that support our analysis techniques to solve large and complex user-role reachability problems. The set of heuristics is classified into two categories: the first category contains heuristics that focus on eliminating as much as possible the number of administrative actions in the original problem that do not contribute to the analysis of ARBAC policy; the heuristics in the second category try to split the original reachability problem into a set of sub-problems that are usually easier to solve by the analysis techniques. An excerpt of an exhaustive experimental evaluation has been conducted and provided evidence that an implementation of the proposed techniques and heuristics, called ASASPXL 2.0, performs significantly better than MOHAWK, VAC, and PMS on a variety of benchmarks from [11,13].

As future work, we plan to design new heuristics based on some functionalities provided by the model checker MCMT such as the capability of tracking the visited states for later use. Another interesting line of research for future work is to consider the combination of backward and forward reachability procedures to speed up the analysis of ARBAC policies. The capability of applying the analysis techniques inside ASASPXL 2.0 to the analysis of location-based privacy policy as discussed in Sect. 5 is also an interesting future work.

Acknowledgement. This research is funded by Vietnam National University Ho Chi Minh City (VNU-HCM) under grant number C2017-20-17.

References

1. http://homes.di.unimi.it/~ghilardi/mcmt
2. http://research.microsoft.com/en-us/um/redmond/projects/z3
3. Alberti, F., Armando, A., Ranise, S.: ASASP: automated symbolic analysis of security policies. In: Björner, N., Sofronie-Stokkermans, V. (eds.) CADE 2011. LNCS (LNAI), vol. 6803, pp. 26–33. Springer, Heidelberg (2011). https://doi.org/10.1007/978-3-642-22438-6_4

4. Alberti, F., Armando, A., Ranise, S.: Efficient symbolic automated analysis of administrative role based access control policies. In: Proceedings of 6th ACM Symposium on Information, Computer and Communications Security (ASIACCS 2011). ACM PR (2011)

5. Ardagna, C.A., Cremonini, M., Vimercati, S.D.C., Samarati, P.: Privacy-enhanced location-based access control. In: Gertz, M., Jajodia, S. (eds.) Handbook of Database Security Applications and Trends, pp. 531–552. Springer, Boston (2008). https://doi.org/10.1007/978-0-387-48533-1_22

6. Armando, A., Ranise, S.: Automated symbolic analysis of ARBAC-policies. In: Cuellar, J., Lopez, J., Barthe, G., Pretschner, A. (eds.) STM 2010. LNCS, vol. 6710, pp. 17–34. Springer, Heidelberg (2011). https://doi.org/10.1007/978-3-642-22444-7_2

7. Bellavista, P., Corradi, A., Giannelli, C.: Efficiently managing location information with privacy requirements in Wi-Fi networks, a middleware approach. In: Proceedings of the Second International Symposium on Wireless Communication Systems, pp. 1–8. IEEE (2005)

8. Crampton, J.: Understanding and developing role-based administrative models. In: Proceedings of 19th ACM Conference on Computer and Communications Security (CCS 2005), pp. 158–167. ACM PR (2005)

9. Cuellar, J.R.: Location information privacy. In: Sarikaya, B. (ed.) Geographic Location in the Internet, pp. 179–208. Kluwer Academic Publishers, Boston (2002)

10. De Capitani di Vimercati, S., Foresti, S., Jajodia, S., Samarati, P.: Access control policies and languages. Int. J. Comput. Sci. Eng. (IJCSE) 3(2), 94–102 (2007)

11. Ferrara, A.L., Madhusudan, P., Nguyen, T.L., Parlato, G.: VAC - verifier of administrative role-based access control policies. In: Biere, A., Bloem, R. (eds.) CAV 2014. LNCS, vol. 8559, pp. 184–191. Springer, Cham (2014). https://doi.org/10.1007/978-3-319-08867-9_12

12. Ghilardi, S., Ranise, S.: Backward reachability of array-based systems by SMT solving: termination and invariant synthesis. Log. Methods Comput. Sci. (LMCS) 6(4), 1–48 (2010)

13. Jayaraman, K., Ganesh, V., Tripunitara, M., Rinard, M., Chapin, S.: Automatic error finding for access-control policies. In: Proceedings of 18th ACM Conference on Computer and Communications Security (CCS 2011). ACM (2011)

14. Jha, S., Li, N., Tripunitara, M.V., Wang, Q., Winsborough, H.: Towards formal verification of role-based access control policies. IEEE Trans. Dependable Secure Comput. 5(4), 242–255 (2008)

15. Li, N., Tripunitara, M.V.: Security analysis in role-based access control. ACM Trans. Inf. Syst. Secur. 9(4), 391–420 (2006)

16. Mohamed, F.M.: Privacy in location-based services: state-of-the-art and research directions. In: Proceedings of the 8th IEEE International Conference on Mobile Data Management (MDM 2007). IEEE (2007)

17. Ranise, S., Truong, A.: Incremental analysis of evolving administrative role based access control policies. In: Atluri, V., Pernul, G. (eds.) DBSec 2014. LNCS, vol. 8566, pp. 260–275. Springer, Heidelberg (2014). https://doi.org/10.1007/978-3-662-43936-4_17

18. Ranise, S., Truong, A., Armando, A.: Boosting model checking to analyse large ARBAC policies. In: Jøsang, A., Samarati, P., Petrocchi, M. (eds.) STM 2012. LNCS, vol. 7783, pp. 273–288. Springer, Heidelberg (2013). https://doi.org/10.1007/978-3-642-38004-4_18

19. Ranise, S., Truong, A., Armando, A.: Scalable and precise automated analysis of administrative temporal role-based access control. In: Proceedings of 19th Symposium on Access control Models and Technologies (SACMAT 2014), pp. 103–114. ACM (2014)

20. Ranise, S., Truong, A., Traverso, R.: Parameterized model checking for security policy analysis. Int. J. Softw. Tools Technol. Transfer (STTT) **18**, 559–573 (2016)

21. Ranise, S., Truong, A., Viganó, L.: Automated analysis of RBAC policies with temporal constraints and static role hierarchies. In: Proceedings of the 30th ACM Symposium on Applied Computing (SAC 2015), pp. 2177–2184. ACM (2015)

22. Sandhu, R., Coyne, E., Feinstein, H., Youmann, C.: Role-based access control models. IEEE Comput. **2**(29), 38–47 (1996)

23. Sasturkar, A., Yang, P., Stoller, S.D., Ramakrishnan, C.R.: Policy analysis for administrative role based access control. In: Proceedings of 19th IEEE Computer Security Foundations Symposium (CSF 2006). IEEE Press, July 2006

24. Stoller, S.D., Yang, P., Ramakrishnan, C.R., Gofman, M.I.: Efficient policy analysis for administrative role based access control. In: Proceedings of 21st ACM Conference on Computer and Communications Security (CCS 2007). ACM Press (2007)

25. Truong, A.T., Dang, T.K., Küng, J.: On guaranteeing k-anonymity in location databases. In: Hameurlain, A., Liddle, S.W., Schewe, K.-D., Zhou, X. (eds.) DEXA 2011. LNCS, vol. 6860, pp. 280–287. Springer, Heidelberg (2011). https://doi.org/10.1007/978-3-642-23088-2_20

26. Truong, A., Hai Ton That, D.: Solving the user-role reachability problem in ARBAC with role hierarchy. In: Proceedings of 2016 International Conference on Advanced Computing and Applications (ACOMP 2016), pp. 3–10. IEEE (2016)

27. Truong, A.T., Truong, Q.C., Dang, T.K.: An adaptive grid-based approach to location privacy preservation. In: Nguyen, N.T., Katarzyniak, R., Chen, S.M. (eds.) Advances in Intelligent Information and Database Systems. SCI, vol. 283, pp. 133–144. Springer, Heidelberg (2010)

28. Truong, Q.C., Truong, A.T., Dang, T.K.: Memorizing algorithm: protecting user privacy using historical information of location based services. Int. J. Mob. Comput. Multimedia Commun. **2**, 65–86 (2010)

29. Yang, P., Gofman, M.I., Stoller, S., Yang, Z.: Policy analysis for administrative role based access control without separate administration. J. Comput. Secur. **23**, 1–9 (2014)

Partitioning-Insensitive Watermarking Approach for Distributed Relational Databases

Sapana Rani$^{(\boxtimes)}$, Dileep Kumar Koshley, and Raju Halder

Indian Institute of Technology Patna, Patna, India
{sapana.pcs13,dileep.pcs15,halder}@iitp.ac.in

Abstract. This paper introduces an efficient watermarking approach for distributed relational databases, which is generic enough to support database outsourcing and hybrid partitioning. Various challenges, like partitioning and distribution of data, existence of replication etc., are addressed effectively by watermarking different partitions using different sub keys and by maintaining a meta-data about the data distribution. Notably, the embedding and detection phases are designed with the aim of making embedded watermarks partitioning-insensitive. That means, database partitioning and its distribution do not disturb any embedded watermark at all. To the best of our knowledge, this is the first proposal on watermarking of distributed relational databases supporting database outsourcing, its partitioning and distribution in a distributed setting.

Keywords: Watermarking · Distributed databases · Security

1 Introduction

Distributed Database System has always been an efficient solution to manage large-scale databases over computer networks [2]. In recent years, adapting cloud-based technology to avail its efficient and cost-effective services are continuously gaining paramount attention from both academia and industry. For instance, cloud-based Database-as-a-Service (DBaaS), such as Amazon Relational Database Service (RDS) [3], Microsoft SQL Azure [4], etc., is now-a-days appealing for organizations to outsource their databases. In particular, a shared platform (e.g., database server hardware and software) is provided to host multiple outsourced databases, leading to a scalable, elastic and economically viable solution. Clients can easily deploy their own databases in the cloud to avail all required services without investing much on resources [5]. In a cloud-based distributed database system, data owners outsource their databases to a cloud-based service provider that partitions and distributes them among multiple servers interconnected by a communication network.

Unfortunately, outsourcing valuable databases to third-party service providers, without taking proper precaution, may increase the possibility of certain typical attacks, such as copyright infringement, data tampering, integrity violations, piracy, illegal redistribution, ownership claiming, forgery, theft,

© Springer-Verlag GmbH Germany 2017
A. Hameurlain et al. (Eds.): TLDKS XXXVI, LNCS 10720, pp. 172–192, 2017.
https://doi.org/10.1007/978-3-662-56266-6_8

etc. [6]. Database watermarking has emerged as a promising technique to countermeasure the above-mentioned threats [6]. This embeds some kind of information (known as watermark) into the data using a secret key, and later extracts the same, on demand, to reason about the suspicious data. For example, suppose a watermark W is embedded into an original database using a private key K which is known only to the owner. On receiving any suspicious database, the owner may perform a verification process using the same private key K by extracting and comparing the embedded watermark (if present) with the original watermark information W.

In the context of centralized database systems, a wide range of works on the watermarking of centralized databases has been proposed over the past decades [7–17]. Specifically, they are based on random bit flipping [8,9], fake tuple insertion [18], random bit insertion [17,19], tuple reordering [13], binary image generation [15,20], generation of local characteristics like range, digit and length frequencies [21], matrix operations [22], etc. In general, they are categorized into *distortion-based* and *distortion-free* depending on whether the embedded watermark distorts the underlying database content or not. These techniques are designed specifically to address the security risks in centralized databases only, making them completely unadaptable to the scenario where databases are partitioned and distributed over a network. As already mentioned, popular applications where people often face such issues include, for example, cloud-based Database-as-a-Service (DBaaS), such as Amazon Relational Database Service (RDS), Microsoft SQL Azure, etc. In general, the use of centralized database watermarking approaches directly in a distributed setting gives rise to various challenges, including (1) distribution of data, (2) existence of replication, (3) preservation of watermarks while performing partitioning and distribution by third party, (4) robustness, (5) efficient key management, etc. These challenges become more prominent when data-partition and data-distribution are done by untrusted third parties (who are different from data owners) as for example in case of cloud-based database as a service model.

To the best of our knowledge, till now there is no significant contribution in case of distributed relational database watermarking, especially for distributed cloud-based database-as-a-service scenarios. Although two related works in this direction are found in [23,24], however they have not considered any core properties of distributed scenario during watermark embedding and detection. Moreover, the proposal does not consider any kind of relational databases and their partitioning over the distributed environment. Authors in [23], although title refers, have not addressed any challenge in distributed database scenario.

Motivating from the above concerns, in [1] we have introduced a preliminary proposal on distributed database watermarking that supports database outsourcing and hybrid partitioning. Although we referred AHK algorithm [9] as partition-level watermarking algorithm, this may have several limitations: AHK algorithm marks only one attribute in a tuple at a time. Therefore, with n vertical partitions, the algorithm should execute the AHK algorithm n times, allowing only one attribute to get marked in a particular partition. In fact, the prime challenge here is to decide which particular part of the tuple will belong to which particular partition.

In this paper[1], we further strengthen the proposal by designing a novel watermarking algorithm which overcome the above mentioned limitations. The watermarking algorithm, in a single execution, can decide to which partition a part of the tuple actually belongs. In other words, re-executing of the complete watermarking algorithm is not required for each vertical partition. Additionally, in a particular partition, we can decide a fraction of attributes (rather than a single attribute) to be marked randomly. The combination of attributes to be marked is decided by secret parameter, increasing the robustness of the approach.

In particular, our main contributions in this paper are:

- Proposing efficient watermarking approach for distributed relational databases, which is generic enough to support database outsourcing and hybrid partitioning.
- Effective treatment to the above-mentioned challenges by watermarking different partitions using different sub-keys and by maintaining a meta-data about the data distribution, without revealing any secret to the third-party.
- Efficient key management using Mignotte's k out of n secret sharing scheme [25], improving the robustness of the scheme.
- The design of embedding and detection phases with the aim of making embedded watermarks partitioning-insensitive. That means, database partitioning and its distribution do not disturb any embedded watermark at all.
- Experimental evaluation on benchmark datasets to establish the effectiveness of our approach in presence of various attacks.

The structure of the paper is as follows: Sect. 2 discusses various works in literature related to database watermarking. Sections 3 and 4 describe the proposed watermark embedding and detection techniques respectively, by illustrating with suitable examples. Analysis on the experimental evaluation results is reported in Sect. 5. A detail comparative study *w.r.t.* the literature is reported in Sect. 6. Finally we conclude our works including the future plans in Sect. 7.

2   Related Works

This section briefly discusses the state-of-the-art on the database watermarking techniques in the literature.

The idea to secure a database of map information (represented as a graph) by digital watermarking technique was first given by Khanna and Zane in 2000 [26]. The first watermarking scheme for relational databases was given by Agrawal et al. in 2002 [9]. They embed the watermark in the least significant bits (LSB) of a particular bit position of some of the selected attribute of some of the selected tuples based on the secret parameters. Started with these pioneer works, a wide range of watermarking techniques for centralized database has been proposed [7–17].

---

[1] This work is a revised and extended version of [1].

Broadly, the existing approaches are categorized into *distortion-based* [8–11,16,17] and *distortion-free* [7,12–15] watermarking techniques, based upon whether distortion occurs or not in the underlying data. The distortion introduced when embedding watermarks should not affect the usability of the data. Authors in [8] have applied data flow analysis to detect variant and invariant part in the database and subsequently watermarked the invariant part. Authors in [10] proposed a reversible-watermarking technique which allows to recover the original data from the distorted watermarked data. Image as watermark is embedded at bit-level in [11]. Approaches in [16,17] are based on database-content - the characteristics of database data is extracted and embedded as watermark into itself. A recent survey by Xie et al. is reported in [27] with special attention to the distortion-based watermarking. Unlike distortion-based techniques, the distortion-free watermarking techniques generate watermark from the database itself. In [13,20], hash value of the database is extracted as watermark information. Approaches in [7,14,15] are based on the conversion of database relation into a binary form to be used as watermark. Authors in [15] used the Abstract Interpretation for verifying integrity of relational databases. A comprehensive survey on various types of watermarks and their characteristics, possible attacks, and the state-of-the-art can be found in [6].

Watermarking schemes can also be classified as *robust* [28,29] and *fragile* [13,17,21,22]. Generally, the digital watermarking for integrity verification is called fragile watermarking as compared to robust watermarking for copyright protection. Recently proposed fragile watermarking techniques include [30,31]. In [30], the proposed technique establishes a one-to-one relationship between the secret watermark and the relative order of tuples in a group. Watermark generation in [31] is based on local characteristics of the relation itself such as frequencies of characters and text length.

Recently in [32], we have proposed watermarking for large scale relational databases by adapting the potential of MapReduce [33] paradigm. The experimental results demonstrated a significant improvement in watermarking cost for distortion free watermarking with respect to the existing sequential algorithms.

To the best of our knowledge, till now there is no significant contribution in case of watermarking of distributed relational database systems. Although two related works in this direction are found in [23,24], however they have not considered any core properties of distributed scenario during watermark embedding and detection. To be more precise, the authors in [24] proposed a real-time watermarking technique for digital contents which are distributed among a group of parties in hierarchical manner. One such example is the distribution of digital works over the Internet involving several participants from content producers to distributors to retailers and finally to customers. The main idea is to perform multilevel watermarking in order to detect attack possibly occurred at any particular level. Unfortunately, their proposal has not considered any kind of relational databases and their partitioning over distributed environment. The major drawback in [24] is that the data owner has to extract all the watermarks from top to bottom in the hierarchy during verification. Authors in [23], although title refers, have not addressed any challenge in distributed database scenario.

3   Watermarking Technique for Distributed Databases

This section proposes a generic watermarking technique for distributed databases. The proposal is based on the scenario where database owner outsources data to a third party, assuming that the third party has the required resources to manage it. Some of the challenges addressed by the proposed technique are: (1) Distribution of data, (2) Existence of replication, (3) Non-disturbance of the embedded watermark during partitioning and distribution, (4) Robustness, etc.

The watermark embedding phase consists of the following steps:

3.1   Step 1: Initial Exchange of Partition Information

Data owner will initiate this process to exchange some basic information with the third party in order to obtain some initial information about the partitioning and distribution of the database.

Let *DB_schema* be a relational database schema. Let *INF* be a set of specifications on the database and its associated applications, which must be preserved after partitioning and distribution by a third party. For example, *INF* may include confidentiality and visibility constraints [34], user access information [2], query behaviours [35], etc. To start this process, the data owner provides *DB_schema* and *INF* to the third party. As a result, the third party will send back to the owner a *partition overview* of the database. This *partition overview* includes information about the set of partitions to be followed in future by the third party.

Let us formalize the *partition overview*: Let *R_schema* be a schema of a database relation belonging to *DB_schema*. The horizontal partitioning of *R_schema* is formally represented by $\langle R\_schema, f_h \rangle$ where f_h is a partial function defined over the set of all attributes A in *R_schema* (i.e. $f_h : A \nrightarrow 2^{\Phi}$ where Φ is the set of all possible well-formed formulas defined on A in first order logic) [36]. In other words, f_h which is expressed in first order predicate formulas on attributes, represents properties of database tuples. The horizontal partitioning of tuples in an instance of *R_schema* is performed based on the satisfiability of f_h. Given $\Phi = \{\phi_1, ..., \phi_m\}$, since there are at most $2^{|\Phi|}$ number of horizontal partitions depending on the satisfiability of predicates Φ, we will follow the following convention: a horizontal partition is represented by h, where h is the decimal conversion of truth values of $\{\phi_1, ..., \phi_m\}$ obtained based on their satisfiability by its tuples. For example, given two properties ϕ_1 and ϕ_2, if a tuple t satisfies $\neg\phi_1 \wedge \neg\phi_2$ then t is assigned to the partition-0 (which is decimal equivalent of truth value "00"). Similarly, if t satisfies $\phi_1 \wedge \neg\phi_2$ then t is assigned to the partition-2 (as the decimal equivalent of truth value "10" is 2) and so on. In the similar way, we also formalize the vertical partitioning as $\langle R\_schema, f_v \rangle$, where $\wp(A)$ is the power set of A and $f_v(A) \subseteq \wp(A)$.

Observe that the definitions of f_h and f_v depend on *INF* in order to satisfy it. Therefore, in general, the hybrid partitioning is formally defined as

$\langle R\_schema, f_h, f_v \rangle$. The partition overview ψ of DB_schema satisfying INF is formally defined as

$$\psi \triangleq \{\langle R\_schema, f_h, f_v \rangle \mid R\_schema \in DB\_schema\}$$

This is worthwhile to mention here that our approach is suitable for static partitioning and infrequent dynamic partitioning [37], where in the later case a re-watermarking is necessary to make the detection partition-independent.

Example 1. Let us illustrate this by a running example. Consider the database relation "*T*" depicted in Table 1.

Table 1. Relation "*T*"

	A_0	A_1	A_2	A_3	A_4	A_5	A_6	A_7	A_8	A_9	A_{10}
t_1	1	123	100	20	15	16	21	35	11	100	15
t_2	2	785	200	29	15	16	28	38	12	150	12
t_3	3	456	300	50	11	160	21	35	22	20	13
t_4	4	320	400	36	155	167	20	35	21	170	14
t_5	5	453	500	40	151	126	27	35	24	160	17

Let us assume that INF consists of the following security specifications:

1. Confidentiality constraints, $C = \{A_2 \wedge A_7 \wedge A_8\} \wedge \{A_3 \wedge A_9 \wedge A_{10}\}$. This means that, none of the partitions should contain either A_2, A_7 and A_8 in combination or A_3, A_9 and A_{10} in combination.
2. Visibility constraints, $V = \{ A_1 \wedge A_2 \wedge A_3\} \vee \{A_6 \wedge A_9 \wedge A_{10}\}$. This means that, there should be at least one partition that contains either A_1, A_2 and A_3 in combination or A_6, A_9 and A_{10} in combination.

Data owner sends T_schema, the schema of relation "*T*", along with the set of specifications INF to the third party. The third party generates the horizontal partitions F_1 and F_2 after applying $f_h : A \rightarrow \phi$ where $A = $ `attribute`$(T\_schema) = \{A_0, A_1, ..., A_{10}\}$ and $\phi : (A_3 \leq $ `average`$(A_3)) \wedge (A_8 \leq $ `average`$(A_8))$. Since we have only one predicate, we can have at most two horizontal partitions F_1 and F_2, where F_1 satisfies ϕ and F_2 doesn't satisfy ϕ. Satisfying INF [34], third party then partitions F_1 and F_2 vertically by applying a suitable function f_v as

$$f_v(F_1) = \{F_{11}, F_{12}\} \qquad and \qquad f_v(F_2) = \{F_{21}, F_{22}\}$$

where

$$F_{11} = \langle \{A_0, A_1, A_2, A_3, A_4, A_5\}, A_3 \leq \text{average}(A_3)\rangle$$
$$F_{12} = \langle \{A_0, A_6, A_7, A_8, A_9, A_{10}\}, A_8 \leq \text{average}(A_8)\rangle$$
$$F_{21} = \langle \{A_0, A_1, A_2, A_3, A_4, A_5\}, A_3 > \text{average}(A_3)\rangle$$
$$F_{22} = \langle \{A_0, A_6, A_7, A_8, A_9, A_{10}\}, A_8 > \text{average}(A_8)\rangle.$$

Finally, the third party sends back this partition overview $\psi = \{F_{11}, F_{12}, F_{21}, F_{22}\}$ to the data owner.

3.2   Step 2: Watermarking by Data Owner

Given a partition overview ψ (provided by the third party) and a secret key K, the data owner embeds watermark into the original database relation R. To this aim, the data owner performs the following two steps:

- *Key Management*: Obtain a set of n different sub-keys $\{K_i \mid i = 1, 2, \ldots, n\}$ from K where n represents the number of partitions obtained from the partition overview ψ (denoted $|\psi|$), and
- *Watermark Embedding*: Embed the watermark W into R using n sub-keys.

Let us describe each step in detail:

Key Management. Since our aim is to make the watermark partitioning-insensitive, the prime challenge here is to select private key K properly and to watermark the database by using K in such a way that partitioning of the database R by third party must not affect the embedded watermark. Watermarking by the data owner considering the future partitioning (yet to be done by third party) leads to following four possibilities:

- Same Watermark, Same Key: Embedding same watermark into different partitions using same key.
- Different Watermark, Same Key: Embedding different watermarks into different partitions using same key.
- Same Watermark, Different Key: Embedding same watermark into different partitions using different keys.
- Different Watermark, Different Key: Embedding different watermarks into different partitions using different keys.

In first and second case, if the watermark is revealed at one site, the key will be exposed and watermarks at other sites also become vulnerable. In the last case, applying different watermarks along with different keys will be a tedious job. Therefore, in our approach, we consider the third scenario, i.e. "Same Watermark, Different Key", in which if somehow the watermark is extracted at one site, it will not expose the watermarks embedded into other database-partitions at other sites. Moreover, this serves the purpose of making watermark detection partition-independent as well. In particular, to achieve our objective, we consider k out of n secret sharing schemes [25,38].

Various k out of n secret sharing schemes are already proposed in the literature such as: Shamir's scheme [38], Mignotte's scheme [25], etc. It states that given a secret K and n shares, any set of k shares acts as the threshold from which the secret can be recovered. In other words, any set of $(k-1)$ shares is not enough to reveal K. However in Shamir's scheme, the attacker gets a range of numbers to guess about the secret key even with $(k-1)$ keys. In our approach, we use Mignotte's scheme as this leads to small and compact shares [39].

Algorithm 1. KEY-COMPUTATION

Input : Partition overview ψ, Secret key K

Output : Shares $\{K_i \mid i = 1, 2, \ldots, n\}$ of the secret key K

1: Let $n = |\psi|$ and k be a threshold, where $|\psi|$ represents the number of partitions.
2: Choose n pairwise co-prime integers $m_1, m_2, \ldots, m_n | (m_1 \times \ldots \times m_k) > (m_{n-k+2} \times \ldots \times m_n)$.
3: Select secret key K such that $\beta < K < \alpha$ where $\alpha = (m_1 \times \ldots \times m_k)$ *and* $\beta = (m_{n-k+2} \times \ldots \times m_n)$.
4: **for** each $i \in 1$ to n **do**
5:      Compute shares of secret key as $K_i = K \bmod m_i$
6: **end for**
7: Return $\{K_i \mid i = 1, 2, \ldots, n\}$.

Algorithm 1 provides detail steps of the Mignotte's scheme to obtain n shares of secret key. Here $n = |\psi|$ indicates the number of partitions. We have a secret key K which is partitioned into different shares, $\{K_i \mid i = 1, 2, \ldots, n\}$ that are used in watermarking of various partitions. Observe that this reduces the challenges in managing and distributing large number of independent keys for all database-partitions in distributed settings.

Example 2. Let us illustrate this using the running example. Consider the partition overview $\psi = \{F_{11}, F_{12}, F_{21}, F_{22}\}$ in Example 1. We require four different keys for watermarking of these four partitions. Considering the threshold k equal to 3, the owner has to assume four pairwise co-prime integers such that the product of k smallest numbers should be greater than the product of $k - 1$ biggest numbers. Suppose they are: $m_1 = 7$, $m_2 = 17$, $m_3 = 3$, $m_4 = 19$. Since $m_1 \times m_2 \times m_3 = 357$ and $m_3 \times m_4 = 57$, it satisfies the condition $m_1 \times m_2 \times m_3 > m_3 \times m_4$. A secret K should be chosen between the range of these two products, let it be $K = 131$. Secret shares are calculated for all n by using $K_i = K \bmod m_i$ as follows:

$$K_1 = K \bmod m_1 = 131 \bmod 7 = 5$$
$$K_2 = K \bmod m_2 = 131 \bmod 17 = 12$$
$$K_3 = K \bmod m_3 = 131 \bmod 3 = 2$$
$$K_4 = K \bmod m_4 = 131 \bmod 19 = 17$$

These set of secret shares or sub-keys K_1, K_2, K_3, K_4 will be used to watermark the relation T at partition-level based on partition overview ψ.

In the rest of the paper, we use the terms "secret share" and "sub-key" synonymously.

Watermark Embedding. In this subsection, we present watermark embedding step by the data owner using partition overview and the set of sub-keys. The primary objective here is to embed same watermark on multiple partitions using different sub-keys. Let us formalize the distributed watermark embedding, as below:

$$\texttt{DistWM\_Embed}(R, \psi, W, K)$$
$$= \bigcup_{i \in 1...|\psi|} \texttt{WM\_Embed}(R^i, W, K_i)$$
$$= \bigcup_{i \in 1...|\psi|} R_w^i$$
$$= R_w$$

Data owner watermarks the database relation R using shares $\{K_i \mid i = 1, 2, ..., |\psi|\}$, obtained from the secret key K in Algorithm 1. Suppose R^i represents i^{th} partition in the partition overview ψ. Observe that R^i is watermarked using the share K_i. Once watermarked, data owner then outsources all the watermarked relations R_w in the database to the third party.

We formalize our watermark embedding algorithm in Algorithm 2. The descriptions of various notations to be used in the algorithms are depicted in Table 2. Observe that all these parameters are secret to the data owner.

Table 2. Notations used in Algorithm 2

Symbol	Description
γ	Fraction of tuples marked during embedding
β	No of bits to be extracted to make the watermark
ℓ	No of attributes available for marking
η	Fraction of attributes to be marked
α	Detectability level
ξ	No. of least significant bit available for marking in an attribute

Algorithm 2 takes database relation R, partition overview ψ and secret key shares (in matrix form) as input, and gives watermarked relation R_w as output. Step 2 in the algorithm checks if a tuple should be marked or not. A tuple t is considered for embedding, if modulus of the hash of t's primary key by γ is zero. A suitable example of hash function is MD5 [40]. Step 3 calls CHECK function to determine the horizontal partition in which a tuple t belongs to. Given a set of predicates $\phi_1, \ldots, \phi_m$, the CHECK function checks which predicates in first order logic are satisfied by t and returns a horizontal partition id h that is the decimal conversion of truth values of $\{\phi_1, ..., \phi_m\}$ based on their satisfiability. After getting the horizontal partition id h, the embedding key $K_{h,v}$ for each vertical partition $\text{VP}_v \in \psi$ is computed at Step 5 from the key matrix $K_{2^{|\phi|} \times |\text{VP}|}$. In Step 6, the owner compress its original watermark, which is to be embedded, in a length of β bits based on the hash-based computation. Finally, the embedding procedure Apply_Watermark is called to watermark the partition by its corresponding sub-key.

Algorithm 2. WM_Embed

Input : Database relation R, Partition overview ψ, Secret keys matrix $K_{2^{|\phi|} \times |VP|}$
Output : Watermarked relation R_w
1: **for** each tuple $t \in R$ **do**
2:       **if** hash$(t.PK)$ mod $\gamma = 0$ **then**   /* $t.PK$ is the primary key of t */
3:             Horizontal partition id $h = $ CHECK$(t, \langle \phi_1, ..., \phi_m \rangle)$
4:             **for** each vertical partition $VP_v \in VP$ **do**
5:                   Watermark key $wm_{key} = K_{h,v}$
6:                   $mark = $ compress$($hash$(owner's\ watermark), \beta)$
7:                   List $L = [t.PK, t.A_x, ..., t.A_y], \forall A_x, ..., A_y \in VP_v$
8:                   Apply_Watermark$(wm_{key}, mark, L)$
9:             **end for**
10:       **end if**
11: **end for**

12: **function** CHECK(TUPLE t, PROPERTIES $\langle \phi_1, ..., \phi_m \rangle)$
13:       Create a boolean array b of length m
14:       **for** $i = 1; i <= m; i = i + 1$ **do**
15:             **if** $t \models \phi_i$ **then**
16:                   $b[i] = 1$
17:             **else**
18:                   $b[i] = 0$
19:             **end if**
20:       **end for**
21:       **return** decimal(b)
22: **end function**

23: **procedure** APPLY_WATERMARK(KEY K, WATERMARK wm, $L = [t.PK, t.A_x, ..., t.A_y])$
24:       no of attributes available for marking, $\ell = |L| - 1$ /* primary key $t.PK$ is unavailable for marking */
25:       no of attributes to be marked, $n = \lceil \ell \times \eta \rceil$ /* η is the fraction of attributes to be marked */
26:       total number of attribute combinations for embedding, $c = \binom{\ell}{n}$
27:       the combination of attributes to be marked, $q = $ hash$(K \parallel t.PK)$ mod c /*each combination contains the list of n attributes */
28:       **for** each attribute $a \in q^{th}$ combination **do**
29:             $bit\_index\ b = hash(K \parallel t.PK)$ mod ξ
30:             Replace b^{th} LSB bit of a with b^{th} bit of wm
31:       **end for**
32: **end procedure**

The Apply_Watermark procedure takes as input a secret key K, watermark wm and the list of attribute values belonging to a particular partition. In Steps 24 and 25, the number of attributes n to be marked is calculated by multiplying the number of attributes ℓ available for watermarking and the predefined fraction η of attributes to be marked. In order to randomize the set of attribute values to be marked for different tuples, Step 26 computes the total number of combination of attributes. For example, if the total number of attributes available for marking (i.e., ℓ) is 5 and we want to mark 30% (i.e., $\eta = 0.3$) of attributes, then the number of attributes to be marked $n = \lceil \ell \times \eta \rceil = 2$. Therefore, total number of combinations each containing 2 attributes out of 5 is $\binom{\ell}{n} = \binom{5}{2} = 10$. Step 27 chooses q^{th} combination, and in Steps 28–31 all attributes in q^{th} combination are marked by replacing their b^{th} least significant bit with the bth least significant bit of the watermark wm, where b is the bit index computed by modulus operation of primary key hash value with the number of least significant bit available for marking, ξ.

A situation may arise where partition size is too small. In such case, the tuples in the partition can be divided into multiple groups and similarly the watermark of larger length can be divided into multiple parts, thus enabling to embed one part in one group of tuples.

Example 3. Consider the running example. Given $\psi = \{F_{11}, F_{12}, F_{21}, F_{22}\}$ and the secret sub keys $K_1 = 5, K_2 = 12, K_3 = 2, K_4 = 17$. These keys are stored in 2×2 matrix $\{\{5, 12\}, \{2, 17\}\}$. Given the partition overview ψ computed in Example 1, its clear that total number of horizontal partition is two and vertical partitions is also two. For tuple t_1, the CHECK function returns $h = 0$ because first tuple doesn't satisfy the predicate $\phi : (A_3 \leq \text{average}(A_3)) \wedge (A_8 \leq \text{average}(A_8))$. Therefore for vertical partition $\{A_0, A_1, A_2, A_3, A_4, A_5\}$, secret key 5 will be used for watermark embedding. Let us assume that the watermark $mark = $ "10" and List $L = [1, 123, 100, 20, 15, 16]$. APPLY_WATERMARK procedure determines the number of attributes available for marking $\ell = |L| - 1 = 6 - 1 = 5$ (A_0 is the primary key and hence unavailable for watermarking). Suppose $\eta = 0.3$, then number of attributes to be marked is $n = \lceil \ell \times \eta \rceil = \lceil 5 \times 0.3 \rceil = 2$. Since we are having 5 attributes, out of which 2 have to be marked, hence the number of possible combinations of 2 attributes is $c = \binom{5}{2} = 10$. Assuming that the 1st combination containing $\langle 1, 2 \rangle$ is chosen by step 27. That is, attributes A_1 and A_2 need to be marked. Let $\xi = 2$ and the *bit_index* b computed for these attributes are 0 and 1 respectively. For embedding the watermark wm, data owner replaces the 0th and 1st LSB bit of A_1 (=123) and A_2 (=100) respectively by the corresponding LSB bit of wm, resulting into 122 and 100 respectively. Similarly another vertical partition $\{A_0, A_6, A_7, A_8, A_9, A_{10}\}$ in t_1 is also marked following the similar steps. This continues for the remaining t_2, t_3, t_4 and t_5.

The watermarked relation is shown in Table 3. Data represented in bold denotes watermarked values. Data owner outsources this watermarked relation to the third party for partitioning and distribution.

Table 3. Watermarked relation "T_w"

	A_0	A_1	A_2	A_3	A_4	A_5	A_6	A_7	A_8	A_9	A_{10}
t_1	1	**122**	**100**	20	15	16	21	**36**	11	**101**	15
t_2	2	785	200	**25**	**14**	16	28	**39**	12	**146**	12
t_3	3	456	**300**	50	**9**	160	21	35	22	**22**	**13**
t_4	4	**322**	**401**	36	155	167	**20**	**34**	22	170	14
t_5	5	**453**	**500**	40	151	126	27	**35**	**27**	160	17

3.3 Step 3: Partitioning and Distribution by Third Party

Once the third party receives the watermarked database relation R_w from the data owner, the third party partitions and distributes it as per the partition overview ψ computed before. A partition may be either a subset of attributes (vertically partitioned) or a subset of tuples with common properties (horizontally partitioned) or both (hybrid partitioning). In addition, the third party maintains a metadata table that contains information about the data distribution over the servers. The metadata information consists of partition ID P_i, property description of the data in the partition in the form of first-order formula, the server ID S_j where partition P_i is located, etc.

Example 4. Considering the partition overview ψ, third party first applies function f_h to horizontally partition "T_w" relation into partitions F_1 and F_2 and then f_v to vertically partition into $F_{11}, F_{12}, F_{21}, F_{22}$. The functions f_h and f_v are the same functions as computed during partition overview creation. The resulting watermarked partitions (shown in Table 4) are finally distributed to different servers. The metadata for our running example is shown in Table 5.

Table 4. Partitioning and distribution by third party

A_0	A_1	A_2	A_3	A_4	A_5
1	**122**	**100**	20	15	16
2	785	200	**25**	**14**	16

(a) F_{11}

A_0	A_6	A_7	A_8	A_9	A_{10}
1	21	**36**	11	**101**	15
2	28	**39**	12	**146**	12

(b) F_{12}

A_0	A_1	A_2	A_3	A_4	A_5
3	456	**300**	50	**9**	160
4	**322**	**401**	36	155	167
5	**453**	**500**	40	151	126

(c) F_{21}

A_0	A_6	A_7	A_8	A_9	A_{10}
3	21	35	22	**22**	13
4	**20**	**34**	22	170	14
5	27	**35**	**27**	160	17

(d) F_{22}

Table 5. Metadata

Partition ID	Partition description		Server ID
	Schema	Properties	
P_1	$\{A_0, A_1, A_2, A_3, A_4, A_5\}$	$A_3 \leq \mathsf{avg}(A_3)$	S_3
P_2	$\{A_0, A_6, A_7, A_8, A_9, A_{10}\}$	$A_8 \leq \mathsf{avg}(A_8)$	S_1
P_3	$\{A_0, A_1, A_2, A_3, A_4, A_5\}$	$A_3 > \mathsf{avg}(A_3)$	S_2
P_4	$\{A_0, A_6, A_7, A_8, A_9, A_{10}\}$	$A_8 > \mathsf{avg}(A_8)$	S_4

4   Partition-Level Watermark Detection

The data owner initiates the detection process if she suspects any of the attacks on her database partitions or on part of it. The main issue here is to know the actual key that was used at the time of watermark embedding. For this purpose, the data owner communicates with the third party to obtain a partition ID P_i and the corresponding server ID S_j based on the matching of the suspicious database data with the property description of P_i in the metadata table.

Algorithm 3. WM_Detect

Input : suspicious partition F, secret key K_i
Output : detection result either as "success" or "fail"
1: total count = match count = 0
2: **for** each tuple $t' \in F$ **do**
3:      **if** hash$(t.PK)$ mod $\gamma = 0$ **then**
4:          total count = total count + 1
5:          $\ell' =$ (no. of attributes in F) - 1   /* $t.PK$ is unavailable for marking */
6:          $n' = \lceil \ell' \times \eta \rceil$ // η is the fraction of attributes marked during embedding
7:          total number of attribute combinations, $c' = \binom{\ell'}{n'}$
8:          $q' =$ hash$(K_i \parallel t.PK)$ mod c' /* each combination contains the list n attributes */
9:          **for** each attribute $a \in q'$ th combination **do**
10:             bit_index $b=$ $hash(K_i \parallel t.PK)$ mod ξ
11:             b^{th} LSB bit of $W' = b^{th}$ LSB bit of a
12:         **end for**
13:         **if** $(W' = W)$ **then**
14:             match count = match count + 1
15:         **end if**
16:     **end if**
17: **end for**
18: threshold $\tau =$ (total count $\times \alpha$)   /* α is the detectability level */
19: **if** match count $\geq \tau$ **then**
20:     verification = "success"
21: **else**
22:     verification = "fail"
23: **end if**

Once the partition ID P_i is obtained, the owner will use the Mignotte's scheme [25] to obtain the key K_i from secret key K and i^{th} co-prime integer m_i. Algorithm 3 formalizes the watermark detection phase. It takes the suspicious partition F and the secret key K_i as the input, and gives as output the reasoning whether verification is successful or not. Step 3 checks whether the tuple was marked at the time of embedding. Step 4 increments the value of total count, i.e. the total number of tuples marked.

Steps 5–12 follow similar steps as in the case of embedding algorithm and identify the marked positions in the tuples to extract the embedded watermark bits. On matching the extracted watermark with the original watermark in step 13, the match count is incremented in the next step. Finally step 19 checks whether the match count crosses the threshold τ, and if so, the watermark detection is considered as successful.

Example 5. In the running example let us consider the partition F_{12} in Table 4 as a suspicious one. The data owner will ask third party for the database partition ID (P_i) and the corresponding server ID (S_j). Third party refers to the metadata in Table 5 to compare the property of suspicious data and replies back with the information $P_i = 2$ and $S_j = S_4$ to data owner. Now the data owner will get the corresponding i^{th} co-prime number m_i (i.e. m_2) and obtains K_i (i.e. K_2) following the Algorithm 1. From the running Example 2, $K_2 = K$ mod $m_2 = 131$ *mod* $17 = 12$. To detect the watermark, the data owner invokes Algorithm 3 passing suspicious partition F_{12} and the key $K_2 = 12$ as inputs. Using secret parameters γ, η, let the data owner computes the marked combination in steps

2–8 as A_7, A_9. From these two attribute values, owner will extract two bits '1' and '0' and reconstruct the watermark $W' = $ "10". As $W = W'$, match-count will be increased by 1. Finally if the number of match-count crosses the threshold, the detection is successful.

As an alternative, watermark can also be detected by the application of query preserving approach [35]. Since all servers are equally likely to be suspect, the owner can send an identical query to all the servers. Based on the responses, data owner assigns probabilities to the servers of being suspicious. Further based on a fixed threshold probability, the data owner will get a set of suspicious server IDs. Now the data owner asks the third party for the database partition IDs (P_i's) corresponding to those suspicious servers. Third party then refers to the metadata and replies with a set of corresponding database partition IDs without accessing the schema and properties stored in it. After getting the database partition IDs, owner will get the corresponding key K_i from Algorithm 1. Owner now hit and try among these keys to extract the watermark from the suspicious partition.

Example 6. After getting the suspicious partition F_{12} in Table 4, data owner generates a query and sends it to all the servers. Based on the responses from the servers, let the probabilities assigned to servers S_1, S_2, S_3, S_4 are 0.6, 0.4, 0.5, 0.8 respectively. Assuming the threshold probability as 0.6, the possible suspects are S_1 and S_4, the data owner asks for the corresponding database partition IDs from third party. Subsequently the third party refers to the metadata and replies back the database partition IDs, i.e. 4 and 2. Now the owner will get keys K_4 and K_2 as follows: $K_4 = K \bmod m_4 = 131 \bmod 19 = 17$, $K_2 = K \bmod m_2 = 131 \bmod 17 = 12$. Owner now hit and try among these keys to extract the watermark from the suspicious partition F_{12}.

5   Experimental Analysis

We have performed experiments on a benchmark dataset, namely Forest Cover Type dataset[2]. This dataset contains 581012 tuples, each having 10 integer attributes, 1 categorical attribute, and 44 boolean attributes. We have added an extra attribute *id* to the dataset which serves as primary key, and used all 10 numerical attributes in our experiments. We implemented our algorithms in Java and executed on the system featured with Intel Core i3 processor (2.50 GHz), Windows Operating System, and 4 GB RAM.

In the beginning, we watermark the original dataset using Algorithm 2 with the secret keys obtained from Mignotte's scheme [25] as depicted in Algorithm 1, by varying the number of partitions (i.e., 2, 4, 6, and 8) obtained by applying both horizontal and vertical partitioning. For all partitions, we have performed the experiments on *Count* = 581012 tuples by taking $\gamma = 50$. The fraction

---

[2] University Of California-Irvine KDD Archive: kdd.ics.uci.edu/databases/covertype/ covertype.html.

of attributes to be marked in a partition, η is taken as 0.5, i.e. 50% of the attributes available for marking are actually marked. Then, we simulate update and deletion attacks on various partitions. In the detection process, we use same set of secret parameters as that of embedding phase. We have taken a fixed detectability level $\alpha = 0.3$ to measure the success detectability, i.e. if match count is greater than threshold τ (30% of the total count) then we consider the detection as successful. Table 6 depicts the watermark embedding results (watermark embedding time in millisecond) for various number of partitions in partition-overview.

Table 6. Results of watermark embedding

No of tuples	$|\psi|$	ℓ	ξ	γ	Total_count	Time (msec)
581012	2	5	3	50	11851	429396
581012	4	5	3	50	11851	381082
581012	6	5	3	50	11851	447483
581012	8	5	3	50	11851	490723

Let us now discuss the watermark detection after update and delete attacks[3]. To start, let us first define detectability rate below:

$$detectability\ rate = (Match\_count/Total\_count) \times 100$$

The watermark detectability rate after update attack is depicted in Fig. 1. This is to be noted that "Total_count" denotes the number of tuples marked during embedding before updation and "Match_count" represents the number of times we are able to extract our embedded watermark successfully from various partitions after update attack. We have taken the results by randomly updating 30%, 60%, 90% and 99% tuples of each partition. We can observe that, setting α to 0.3, even after updating 99%, we are able to detect our embedded watermark in some cases. However, owner can tune this value to make a trade off between false positives and false negatives.

Similarly the watermark detection rate after delete attack is shown in Fig. 2. The results have been taken by randomly deleting 30%, 60%, 90% and 99% tuples of each partition.

The average detection times for all partitions after update and delete attacks are shown in Figs. 3 and 4 respectively. Figures 5, 6, 7 and 8 represent the detection rate for 2, 4, 6 and 8 partitions respectively.

Based on these experimental results, we establish the following observations:

- The watermark is successfully detected even after the attacker updates 99% of the tuples in some cases, considering $\alpha = 0.3$. Therefore, detection success will be increased as we decrease α.

---

[3] Detail experimental data can be found in [41].

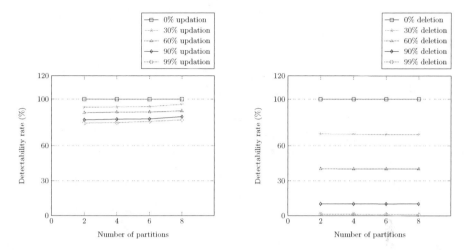

Fig. 1. Watermark detection rate after update attack

Fig. 2. Watermark detection rate after delete attack

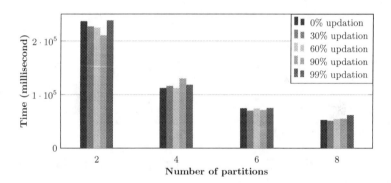

Fig. 3. Average detection time after update attack

- For 0% updation or deletion, we have 100% detectability rate for all partitions, which is always true.
- Figures 1 and 2 show that detectability rate increases as we increase the number of partitions.
- Figures 1 and 2 show that detectability rate decreases as we increase the percentage of updation and deletion.
- It is clear from Figs. 3 and 4 that for the same percentage of updation or deletion, the average detection time decreases as we increase the number of partitions.
- According to Figs. 5, 6, 7 and 8, we observe that the detectability rate decreases on increasing the percentage of updation and deletion.

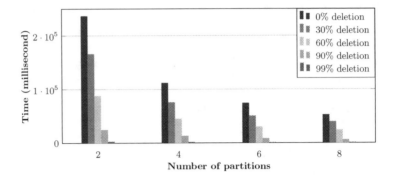

Fig. 4. Average detection time after delete attack

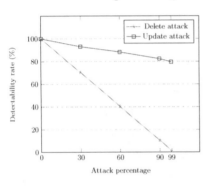

Fig. 5. Percentage of detection for 2 partition

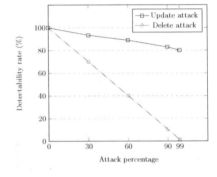

Fig. 6. Percentage of detection for 4 partition

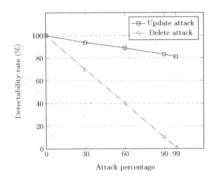

Fig. 7. Percentage of detection for 6 partition

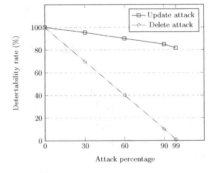

Fig. 8. Percentage of detection for 8 partition

6   Discussions *w.r.t.* the Literature

A wide range of watermarking techniques [7–17] for centralized database has been proposed. Unlike all these, our scenario is based on the cloud-based distributed database system where data owners outsource their databases to a cloud-based service provider who eventually partitions and distributes them among multiple servers interconnected by a communication network.

In the context of distributed database watermarking, let us briefly describe two existing approaches found in the literature:

El-Bakry et al. [23]. The proposed technique, for the purpose of watermarking, changes the structure of relational database schema by adding a new record. This new record is generated by applying a secret function on the original data of each field with the help of a secret key. Though the title refers, they have not addressed any challenge in distributed database scenario. In fact, the major technical contributions have not considered any distributed database scenario at all.

Razdan et al. [24]. The watermarking technique is proposed for digital contents which are distributed among a group of parties in hierarchical manner. For example, distribution of digital works over the internet involving several participants from content producers to distributors to retailers and finally to customers. The proposed technique inserts unique watermarks at each transaction stage to provide a complete audit-trail. This hierarchical watermarking of digital contents imposes difficulties during the watermark extraction process as the data owner has to extract all the watermarks from top to bottom.

This is worthwhile to mention here that the authors in [23,24], however, have not considered any core properties of distributed scenario during watermark embedding and detection. Their proposals do not even consider any kind of database partitioning over the distributed environment.

Advantages of our approach: In this paper, we have proposed a partition-independent database watermarking approach in distributed environment. Since we have embedded same watermark in all the database partitions using different key, the vulnerability of different partitions is minimized. Even if somehow the watermark at one partition is revealed, it will not affect the watermarks embedded in other partitions. The watermark detection can be done on each partition independently. For the purpose of key management we have used k-out of-n secret sharing algorithm [25] which makes our watermark more robust.

Disadvantages of our approach: The initial exchange of partition information between the data owner and the third party service provider induces an overhead in our proposal. The proposed framework relies on the assumption that the service provider always agrees to the previously computed partition-overview at a later stage of partitioning and distribution.

Despite these overheads, our watermarking approach best suites in the cloud computing scenario where data owners outsource their watermarked databases to the third party service providers. The key management scheme makes the detection process partition independent and also an attack at one partition doesn't affect the other partitions at all. The experimental analysis shows that detectability rate increases as we increase the number of partitions since the match count also increases. As obvious, the detectability rate decreases as we increase the percentage of updation and deletion. For 0% updation or deletion, we have 100% detectability rate for all partitions, which is always true.

7   Conclusions and Future Plan

In this paper, we proposed a novel watermarking technique for distributed databases that supports hybrid partitioning. The algorithms are designed to be partitioning-insensitive. The key management scheme that we have considered makes the watermark more robust against various attacks, as if anyhow some partitions are attacked it will not affect any watermark in other database-partitions. The experimental results show the strength of our approach by analyzing the detection rate with respect to random modification and deletion attack. To the best of our knowledge, this is the first work on watermarking of databases in distributed setting that supports database outsourcing and its partitioning and distribution. The future work aims to extend it to the case of big data in cloud computing environment.

Acknowledgment. This work is partially supported by the Council of Scientific and Industrial Research (CSIR), India. We thank the anonymous reviewers for their useful comments and remarks.

References

1. Rani, S., Koshley, D.K., Halder, R.: A watermarking framework for outsourced and distributed relational databases. In: Dang, T.K., Wagner, R., Küng, J., Thoai, N., Takizawa, M., Neuhold, E. (eds.) FDSE 2016. LNCS, vol. 10018, pp. 175–188. Springer, Cham (2016). https://doi.org/10.1007/978-3-319-48057-2_12
2. Özsu, M.T., Valduriez, P.: Principles of Distributed Database Systems. Springer, New York (2011)
3. Amazon Relational Database Service. https://aws.amazon.com/rds/
4. Microsoft Azure SQL Database. https://azure.microsoft.com/en-in/services/sql-database/
5. Curino, C., Jones, E.P., Popa, R.A., Malviya, N., Wu, E., Madden, S., Balakrishnan, H., Zeldovich, N.: Relational cloud: A database-as-a-service for the cloud (2011)
6. Halder, R., Pal, S., Cortesi, A.: Watermarking techniques for relational databases: Survey, classification and comparison. J. UCS **16**(21), 3164–3190 (2010)
7. Bhattacharya, S., Cortesi, A.: A generic distortion free watermarking technique for relational databases. In: Prakash, A., Sen Gupta, I. (eds.) ICISS 2009. LNCS, vol. 5905, pp. 252–264. Springer, Heidelberg (2009). https://doi.org/10.1007/978-3-642-10772-6_19
8. Rani, S., Kachhap, P., Halder, R.: Data-flow analysis-based approach of database watermarking. In: Chaki, R., Cortesi, A., Saeed, K., Chaki, N. (eds.) Advanced Computing and Systems for Security. AISC, vol. 396, pp. 153–171. Springer, New Delhi (2016). https://doi.org/10.1007/978-81-322-2653-6_11

9. Agrawal, R., Haas, P.J., Kiernan, J.: Watermarking relational data: framework, algorithms and analysis. VLDB J. **12**(2), 157–169 (2003)
10. Gupta, G., Pieprzyk, J.: Database relation watermarking resilient against secondary watermarking attacks. In: Prakash, A., Sen Gupta, I. (eds.) ICISS 2009. LNCS, vol. 5905, pp. 222–236. Springer, Heidelberg (2009). https://doi.org/10.1007/978-3-642-10772-6_17
11. Zhou, X., Huang, M., Peng, Z.: An additive-attack-proof watermarking mechanism for databases' copyrights protection using image. In: Proceedings of the 2007 ACM Symposium on Applied Computing, pp. 254–258. ACM (2007)
12. Halder, R., Cortesi, A.: A persistent public watermarking of relational databases. In: Jha, S., Mathuria, A. (eds.) ICISS 2010. LNCS, vol. 6503, pp. 216–230. Springer, Heidelberg (2010). https://doi.org/10.1007/978-3-642-17714-9_16
13. Li, Y., Guo, H., Jajodia, S.: Tamper detection and localization for categorical data using fragile watermarks. In: Proceedings of the 4th ACM workshop on Digital rights management, pp. 73–82. ACM (2004)
14. Li, Y., Deng, R.H.: Publicly verifiable ownership protection for relational databases. In: Proceedings of the 2006 ACM Symposium on Information, Computer and Communications Security, pp. 78–89. ACM (2006)
15. Bhattacharya, S., Cortesi, A.: Distortion-free authentication watermarking. In: Cordeiro, J., Virvou, M., Shishkov, B. (eds.) ICSOFT 2010. CCIS, vol. 170, pp. 205–219. Springer, Heidelberg (2013). https://doi.org/10.1007/978-3-642-29578-2_13
16. Zhang, Y., Niu, X., Zhao, D., Li, J., Liu, S.: Relational databases watermark technique based on content characteristic. In: 2006 First International Conference on Innovative Computing, Information and Control, ICICIC 2006, vol. 3, pp. 677–680. IEEE (2006)
17. Guo, H., Li, Y., Liu, A., Jajodia, S.: A fragile watermarking scheme for detecting malicious modifications of database relations. Inf. Sci. **176**(10), 1350–1378 (2006)
18. Pournaghshband, V.: A new watermarking approach for relational data. In: Proceedings of the 46th Annual Southeast Regional Conference on XX, pp. 127–131. ACM (2008)
19. Kamran, M., Suhail, S., Farooq, M.: A robust, distortion minimizing technique for watermarking relational databases using once-for-all usability constraints. IEEE Trans. Knowl. Data Eng. **25**(12), 2694–2707 (2013)
20. Bhattacharya, S., Cortesi, A.: A distortion free watermark framework for relational databases. In: ICSOFT, vol. 2, pp. 229–234 (2009)
21. Khan, A., Husain, S.A.: A fragile zero watermarking scheme to detect and characterize malicious modifications in database relations. Sci. World J. **2013**, 1–16 (2013)
22. Camara, L., Li, J., Li, R., Xie, W.: Distortion-free watermarking approach for relational database integrity checking. Math. Probl. Eng. **2014**, 1–10 (2014)
23. El-Bakry, H., Hamada, M.: A developed watermark technique for distributed database security. In: Herrero, Á., Corchado, E., Redondo, C., Alonso, Á. (eds.) Computational Intelligence in Security for Information Systems. AISC, vol. 85, pp. 173–180. Springer, Heidelberg (2010)
24. Razdan, R.: Real-time, distributed, transactional, hybrid watermarking method to provide traceability and copyright protection of digital content in peer-to-peer networks, 7 March 2001. US Patent App. 09/799,509
25. Mignotte, M.: How to share a secret. In: Beth, T. (ed.) EUROCRYPT 1982. LNCS, vol. 149, pp. 371–375. Springer, Heidelberg (1983). https://doi.org/10.1007/3-540-39466-4_27

26. Khanna, S., Zane, F.: Watermarking maps: hiding information in structured data. In: Proceedings of the Eleventh Annual ACM-SIAM Symposium on Discrete Algorithms, Society for Industrial and Applied Mathematics, pp. 596–605 (2000)
27. Xie, M.R., Wu, C.C., Shen, J.J., Hwang, M.S.: A survey of data distortion watermarking relational databases. Int. J. Netw. Secur. **18**(6), 1022–1033 (2016)
28. Khanduja, V., Chakraverty, S., Verma, O.P., Singh, N.: A scheme for robust biometric watermarking in web databases for ownership proof with identification. In: Ślęzak, D., Schaefer, G., Vuong, S.T., Kim, Y.-S. (eds.) AMT 2014. LNCS, vol. 8610, pp. 212–225. Springer, Cham (2014). https://doi.org/10.1007/978-3-319-09912-5_18
29. Khanduja, V., Verma, O.P., Chakraverty, S.: Watermarking relational databases using bacterial foraging algorithm. Multimed. Tools Appl. **74**(3), 813–839 (2015)
30. Kamel, I., AlaaEddin, M., Yaqub, W., Kamel, K.: Distortion-free fragile watermark for relational databases. Int. J. Big Data Intell. **3**(3), 190–201 (2016)
31. Alfagi, A.S., Manaf, A.A., Hamida, B., Olanrewajub, R.: A zero-distortion fragile watermarking scheme to detect and localize malicious modifications in textual database relations. J. Theoret. Appl. Inf. Technol. **84**(3), 404–413 (2016)
32. Rani, S., Koshley, D.K., Halder, R.: Adapting mapreduce for efficient watermarking of large relational dataset. In: Trustcom/BigDataSE/ICESS, 2017 IEEE, pp. 729–736. IEEE (2017)
33. Dean, J., Ghemawat, S.: Mapreduce: simplified data processing on large clusters. Commun. ACM **51**(1), 107–113 (2008)
34. De Capitani di Vimercati, S., Foresti, S., Jajodia, S., Paraboschi, S., Samarati, P.: Fragments and loose associations: Respecting privacy in data publishing. In: Proceedings of the VLDB Endowment, vol. 3(1–2), pp. 1370–1381 (2010)
35. Agrawal, S., Narasayya, V., Yang, B.: Integrating vertical and horizontal partitioning into automated physical database design. In: Proceedings of the 2004 ACM SIGMOD International Conference on Management of Data, pp. 359–370. ACM (2004)
36. Huth, M., Ryan, M.: Logic in Computer Science: Modelling and Reasoning about Systems. Cambridge University Press, Cambridge (2004)
37. Rodríguez, L., Li, X.: A dynamic vertical partitioning approach for distributed database system. In: 2011 IEEE International Conference on Systems, Man, and Cybernetics (SMC), pp. 1853–1858. IEEE (2011)
38. Shamir, A.: How to share a secret. Commun. ACM **22**(11), 612–613 (1979)
39. Iftene, S.: General secret sharing based on the chinese remainder theorem with applications in e-voting. Electron. Notes Theoret. Comput. Sci. **186**, 67–84 (2007)
40. Schneier, B.: Applied Cryptography: Protocols, Algorithms, and Source Code in C. Wiley, New York (2007)
41. Rani, S., Kumar Koshley, D., Halder, R.: Partitioning-insensitive watermarking approach for distributed relational databases. Technical report, Department of Computer Science and Engineering, Indian Institute of Technology Patna (2017). http://www.iitp.ac.in/~halder/Papers/TechReport/DistributedRDBWatermark.pdf

Erratum to: Transactions on Large-Scale Data- and Knowledge-Centered Systems XXXVI

Abdelkader Hameurlain[1]([⋈]), Josef Küng[2], Roland Wagner[2],
Tran Khanh Dang[3], and Nam Thoai[3]

[1] IRIT, Paul Sabatier University, Toulouse, France
hameur@irit.fr
[2] FAW, University of Linz, Linz, Austria
[3] Ho Chi Minh City University of Technology, Ho Chi Minh City, Vietnam

Erratum to:
A. Hameurlain et al. (Eds.):
Transactions on Large-Scale Data- and Knowledge-Centered
Systems XXXVI, LNCS 10720,
https://doi.org/10.1007/978-3-662-56266-6

The subtitle "Special Issue on Data and Security Engineering" was omitted from an earlier version of the cover and frontmatter pages of this publication.

The updated online version of this book can be found at
https://doi.org/10.1007/978-3-662-56266-6

© Springer-Verlag GmbH Germany 2018
A. Hameurlain et al. (Eds.): TLDKS XXXVI, LNCS 10720, p. E1, 2017.
https://doi.org/10.1007/978-3-662-56266-6_9

Author Index